THE
ARSONIST

ALSO BY CHLOE HOOPER

A Child's Book of True Crime
The Tall Man
The Engagement

THE
ARSONIST

CHLOE HOOPER

SCRIBNER

LONDON NEW YORK SYDNEY TORONTO NEW DELHI

First published by Penguin Random House Australia Pty Ltd, 2018

First published in Great Britain by Scribner, an imprint of
Simon & Schuster UK Ltd, 2019
This paperback edition published by Scribner, an imprint of
Simon & Schuster UK Ltd, 2020

1 3 5 7 9 10 8 6 4 2

Simon & Schuster UK Ltd
1st Floor
222 Gray's Inn Road
London WC1X 8HB

Simon & Schuster Australia, Sydney
Simon & Schuster India, New Delhi

www.simonandschuster.co.uk
www.simonandschuster.com.au
www.simonandschuster.co.in

A CIP catalogue record for this book is available from the British Library

Paperback ISBN: 978-1-4711-8223-5
eBook ISBN: 978-1-4711-8224-2

Printed and bound by CPI Group (UK) Ltd, Croydon, CR0 4YY

FOR DON

contents

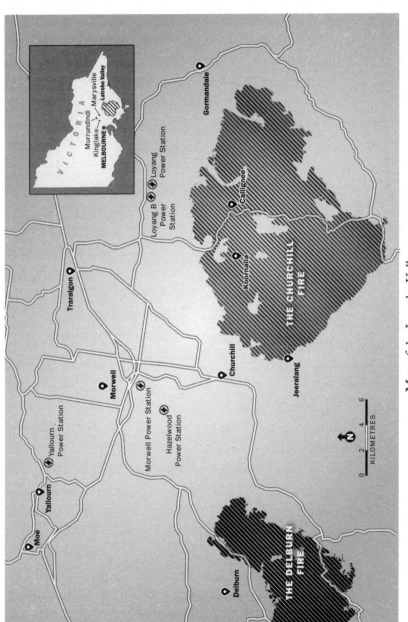

Map of the Latrobe Valley

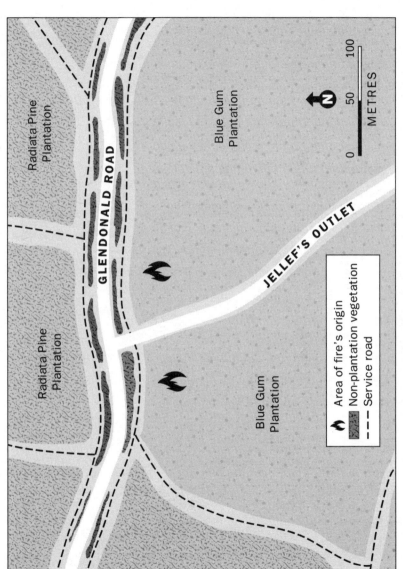

Map of the fire's origins

Part I

the detectives

Picture a fairytale's engraving. Straight black trees stretching in perfect symmetry to their vanishing point, the ground covered in thick white snow. Woods are dangerous places in such stories, things are not as they seem. Here, too, in this timber plantation, menace lingers. The blackened trees smoulder. Smoke creeps around their charcoal trunks and charred leaves. The snow, stained pale grey, is ash. Place your foot unwisely and it might slip through and burn. These woods are cordoned off with crime scene tape and guarded by uniformed police officers.

At the intersection of two nondescript roads, Detective Sergeant Adam Henry sits in his car taking in a puzzle. On one side of Glendonald Road, the timber plantation is untouched: pristine *Pinus radiata*, all sown at the same time, growing in immaculate green lines. On the other side, near where the road forms a T with a track named Jellef's Outlet, stand rows of *Eucalyptus globulus*, the common blue gum cultivated the world over to make printer paper. All torched, as far as the eye can see. On Saturday 7 February 2009, around 1.30 pm, a fire started somewhere near here and now, late on Sunday afternoon, it is still burning several kilometres away.

Detective Henry has a new baby, his first, a week out of hospital. The night before, he had been called back from paternity leave for a 6 am meeting. Everyone in the Victoria Police Arson and Explosives Squad was called back. The past several days had been implausibly hot, with Saturday the endgame – mid-forties Celsius, culminating in a killer hundred-kmh northerly wind. That afternoon and throughout the night, firestorms ravaged areas to the state's north, north-west, north-east, south-east and south-west. Henry was sent two hours east of Melbourne to supervise the investigation of this fire that started four kilometres from the town of Churchill (pop. 4000). An investigation named, for obvious reasons, Operation Winston.

Through the smoke, and in the added haze of the sleep-deprived, he drove with a colleague along the M1 to the Latrobe Valley. On the radio, the death toll was rising – fifty people, then a hundred. Whole towns, it was reported, had burned to the ground. The officers hit the first roadblock an hour out of the city. The dense forest of the Bunyip State Park was on fire, and the traffic police ushered them past onto a ghost freeway. For the next hour they might have been the only car on the usually manic road.

Outside, a string of towns nestled in the rolling green farms of Gippsland, and then it changed to coal country. Latticed electricity pylons multiplied closer to their source, their wires forming waves over the hills.

Turning a corner beyond Moe, Henry saw the cooling towers and cumulus vapours of the first power station, then, round the next bend, a valley ruled by the eight colossal chimneystacks of another station called Hazelwood. A vast open-cut coalmine abutted the highway. Layers of sloping roads descended deep into a brown core – the carbon remnants of a 30-million-year-old swamp – where dredgers, shrunk by a trick of the eye to Matchbox versions, relentlessly gouged the earth.

He turned off to Churchill, a few kilometres south of the highway. The town, built in the late 1960s as a dormitory suburb for electricity workers, had wide streets and a slender, anodised statue rising thirty metres out of the ground. It was the sole public monument, commemorating the great man of Empire in the form of a stylised golden cigar.

The detective didn't stop. He could see smoke above the blackened hills circling the town and wanted to get to the fire's suspected area of origin before it was disturbed. If this was a case of arson, the police needed to prove the connection between the point of ignition and the victims, some of whom were likely to be kilometres away in places still too dangerous to access.

Passing the final roadblock, Henry parked and sat looking at the Nordic dreamscape on one side of the road, and the blackness on the other – the axis where the world had tilted.

Out of the car, it was eerily quiet. No birds cried, no insects thrummed their white noise. The air was cool, pungent with eucalyptus smoke. A not unpleasant smell. On the other side of the police tape, Henry saw the police arson chemist.

George Xydias had slightly hunched shoulders and a slant to his neck as if from his many years looking for clues in ashes and rubble. He had investigated accidental fires and deliberate fires; explosions in cars, boats, trucks, planes; and, after the terrorist attacks of 2002, nightclubs in Bali. He had been to so many scorched crime scenes he could smell what type of vegetation or building material had just been incinerated, and even – to the irritation of those in his laboratory, exposed by his meticulous ways – the percentage of evaporated fuel sometimes left behind.

Wearing disposable white suits, Xydias and his assistant were talking with Ross Pridgeon, a bespectacled, shy, dryly humorous man with a mop of shaggy brown hair. Pridgeon, a local wildfire investigator from the Department of Sustainability and Environment, had been the first to examine the scene that morning. Amongst the precise rows of smouldering blue gums, he'd found signs of two deliberately lit fires, a hundred metres apart on either side of Jellef's Outlet.

Pridgeon showed the assembled police team how he'd traced his way to the place where the two fires joined. Three hundred metres along the outlet, flames had crossed from the east side to the west, high up, flashing in the tops of the trees. This was where a head fire – one of the two burgeoning fire fronts – had come surging through. The eucalypt crowns had been stripped out and the remaining blackened leaves appeared stiff, snap-dried, arrowing in the direction of Saturday's wind. The gum leaves, pliable up to a certain temperature, were like thousands of fingers pointing the way the fire had gone: a sign to the investigators that if they entered the fire zone here and moved back in the opposite direction, they might come to where it started.

The ash-covered ground crunched underfoot. Henry trod carefully to minimise disturbing what could be evidence. He was thirty-six years old, handsome, with an athlete's build and sinewy walk. He'd been selected to play football professionally but a bad injury changed his course. Later, as a detective in the transit unit, dealing with rapes and assaults on public transport, he'd one day been called to a train station that someone had set alight. Standing in the seared crime scene, Henry was impressed that the Arson Squad uncovered a chain of logic in the ash. He applied for a transfer and spent the next few years learning what fire could do, and what answers might be found in its ruins.

This blaze had been so intense the aftermath was like stepping into a textbook on wildfire investigation. Heat was rising off the burnt trees and smoke hung low around the boughs. Henry and the others navigated their way through this smoke, following the subtlest of signs.

Fire is a strange craftsman. It can bevel branches, blunting the wood on the origin side and tapering the back as it advances; it can 'crocodile' a tree's bough, leaving charcoal scales on the point of impact. White ash is the hallmark of complete combustion, and objects directly hit may appear lighter; a fence the men came upon had more of this pale soot on one side, and they moved further that way. Rocks and larger tree limbs often shield finer fuels, such as twigs: where the latter were unburnt, the men knew to move in the opposite direction. They looked for how deeply the timber was charred, and also at the angle of char, another sign of which way the fire had travelled – the scorch pattern on a tree trunk facing the fire's origin was low, whereas there was a steep angle to the burn mark on the sides and back of the trunk as the flames leapt forward.

Now the investigators started walking sideways across the path of the head fire until they found indications of the fire's flank. On the periphery of the blaze, trees were not as badly burnt: fuels that the fire's centre would have destroyed were sometimes barely touched. The men crossed back again and found the fire's other flank. Zigzagging back and forth, narrowing in slowly, they followed a natural V-shape to its point and came to what is known as the area of confidence. Here, paradoxically, the signs were more bewildering. The leaves did not all angle in the same direction: in its coiling infancy, the fire hadn't yet established its course. The damage was closer to the ground. Objects had burned in jagged ways. Somewhere very near here the fire had begun.

Beyond this area were clear signs of the backing fire, or heel fire, where the fledgling flames had reared back, trying to spread until corralled by the wind. Signs of burning were less powerful: fine fuels still remained, and the angle of char was even and level. The investigators started putting flags up to mark the outline of where it seemed the fire had been lit.

Some 26 000 hectares of plantation, state forest and private property had been burnt and yet, after an hour studying and photographing the evidence, the men could tighten their flags into an area of eight square metres, four metres inside the plantation. There was no sign of an incendiary device – sometimes investigators would find the remnants of DIY contraptions made with mosquito coils or party sparklers attached to weights – but in the explosive conditions of the day before, all the arsonist would have needed was a lighter. One flick of the finger and the spark wheel releases terror.

Then again. The second fire had started only a short walk from the first.

A local police officer had found Ross Pridgeon earlier in the day and told him the initial crew attending the blaze had seen two parallel fires burning. Pridgeon led Henry and the other investigators to an area on the western side of Jellef's Outlet, also a few metres in from Glendonald Road. Here too they identified the head fire then crisscrossed the flanks, marking the periphery, working backwards to the area of origin.

This second fire appeared to have started just behind a sign reading PROHIBITION AGAINST DUMPING, regarded locally as an invitation to unload rubbish. There were three bicycles, or the twisted remains of their frames, alongside the burnt debris of old tyres and other car junk, televisions, mattresses, couches, a pram, children's toys – the domestic excess of people unwilling or unable to pay fees at a tip.

None of it was the kind of rubbish that could self-ignite. The investigators looked for signs of glass bottles, which, like a magnifying glass in the hot sun, might kindle dry grass – there were none. There were no junk food containers, or porn, or aerosol cans left by kids chroming – sometimes after getting high they messed around with matches in the woods. There had been no lightning strikes, no heavy machinery nearby; no powerlines were down, and no one would have camped here.

Could an ember from the first fire have created the second? Xydias believed that such 'spotting' was virtually impossible within the first fifteen or twenty minutes of ignition. An ember would have had to travel backwards into blasting wind, then sideways, to light up the other area. The evidence suggested that two high-intensity head fires had moved rapidly south-east, fan-forced by the hot, strong north-westerly. They'd been separately lit, in conditions ideal for a monster blaze.

Twelve years of drought had turned the logs in the plantation's undergrowth, the leaf litter, even the organic matter in the soil, into fuel. The arsonist had had no need to set kindling amongst the blue gums. Each tree had made its own pyre. Every summer they dropped their bark and branches and leaves, and each year without fire the piles grew higher, and they released toxins to ward off new growth that would compromise their fuel beds. No plant on the planet craves fire like the eucalypt: to live it *needs* to burn. 'Gasoline trees', the Americans call the *globulus*. Flames release gases that act like propellant, sending fireballs rolling across treetops. The shedding ribbon bark unfurls streamers of fire that travel kilometres on the wind.

Indigenous Australians managed this pyrophile ecology to their own advantage. Among European settlers it created a sub-community of destructive fire-setters. For generations this had been a kind of open secret. In many country towns there was *someone* who seemed to go on a spree each summer, just as the north winds blew in from the Central Desert. Only relatively recently had the Latrobe Valley been declared a 'hot zone', due to the high rate of deliberately lit fires. Here, it was as if this preference for flames was as much in the DNA of certain locals as it was in the plants.

Ross Pridgeon had also spent much of his previous weekend chasing the work of an arsonist. For two days, detectives and fire investigators had arrived at a series of blazes half an hour after the fire-lighter. Pridgeon would get to one and receive a pager message about the next. The temperature hit forty-five degrees. Soon eight fires had been lit on the outskirts of bushland around the area of Delburn, twenty kilometres west of Churchill. Three of the blazes joined together to make a major fire, destroying forty-four houses and burning 6500 hectares of mainly state forest.

With horror conditions forecast for 7 February, HVP – the timber company that owned this stretch of industrial forest – had their officers patrolling the area, and the police had been surveilling the worst of the region's known firebugs. Despite this, Pridgeon and Xydias were once again taking crime scene photographs of leaf freeze and char patterns. They were checking the ground for, say, the buried head of a matchstick poking from the ash. The fire scientists weren't about to speculate on who had lit this fire. When they turned up to a job, they didn't want to know the local rumours about Firebug X or Y. Nothing but the uncannily expressive evidence concerned them.

Henry's job, however, was just beginning. It is estimated that only one per cent of bushfire arsonists are ever caught. As he got closer to the site of the first flame, it felt like he moved further away. The sudden chaos of the indicators is why the area of confidence is also called the area of confusion. Here, at the site of the first moments of ignition, the evidence reveals the morphing power of what is to come.

In the next half-hour, Henry would drive with Xydias a kilometre up Glendonald Road and, in the rubble that had been a house, find the burnt remains of two brothers. *For I have eaten ashes like bread, and mingled my drink with weeping.* Psalm 102 was something else the Marist Brothers hadn't managed to teach him. He would organise for a police guard to stay on site until the rest of the specialist forensic team arrived on Monday morning. Right now, although it wouldn't get dark for another couple of hours, it was too perilous to drive much further into the wreckage. Roads were still lined with burning trees, their branches waiting to fall.

Instead, Henry planned to check in at the region's police headquarters, ten minutes away in Morwell, and after that, crash for the night at the local motel. He would live here for the next few months. Little plastic-wrapped soaps, long-life milk in the bar fridge, a photo of his new daughter on his phone – the world made miniature, ordered and secure.

A century ago, Henry Lawson wrote that arson expresses a malice 'terrifying to those who have seen what it is capable of. You never know when you are safe.'

As the scientists inspected the ground for signs of whatever the arsonist had used to start the fire, Adam Henry stood in the area of confusion and wondered, *Why?* He'd been trained to think of motivation first – why, then who. So, was this a vendetta? Was it random? Did the arsonist live nearby? Or did someone whom the arsonist was targeting? Was it an act of revenge? Why the plantation? Plenty of environmentalists had protested the slow death of the forest; the privatisation of the Strzelecki Ranges had seen much of what remained of the old-growth mountain ash and blackwood cleared for monocultures of pine and blue gum. Was it for the thrill, the power? Was it psychosis?

Surely whoever did this had known that on such a day a blaze would likely cremate everything in sight? Or was knowing this the reason?

The scientists were not the kind to anthropomorphise. And yet they did. 'Flank', 'head', 'back' or 'rear', 'fingers of flame', 'tongue', 'tail': despite themselves they described a beast. The low-hanging smoke drifted around the burnt trees. A sprite may as well have visited the forest and left one tiny spark, one curling lick of flame that begat this monster, which grew a tongue, a head, flanks and claw-like fingers, and stretched for mile after mile, taking whatever it wanted.

Who, and why?

The patient had been in a coma for twelve days. Strange dreams were all he could remember. He dreamt he was in a red room, then a green room, and when, finally, he woke, the walls were orange. There was flame even in the paint colour and he knew without being told that his wife was dead. He checked his hands and was surprised to find that his fingers – put back together now, bandaged – had been saved.

His children sat next to his bed, while a young police officer had positioned his chair further away, towards the back of the hospital room. All of them were waiting to hear what had happened.

Detective Senior Constable Paul Bertoncello had visited before. The first time, Rodney Leatham had been wearing an oxygen mask and couldn't speak. He had burns to 40 per cent of his body and was covered in layers of dressings. The detective had cautiously touched a bandaged hand. He'd been a police officer for nearly ten years, but no grave situation ever felt like preparation for the next. Rodney was crying, nodding, communicating with his eyes. He was two weeks behind everybody else: already, even by the tiniest increments, people were adapting to the reality of the fire, but he was still right at the beginning.

'I know it's hard,' the detective found himself saying, 'but I'll need to come back and talk a bit.'

Again and again, he bypassed the foyer's balloons and flowers, and took the elevator to the burns unit. On the ward there were nineteen people who'd survived the fires. They had submerged themselves in any liquid they could find, in dams or livestock troughs, which saved their lives but left many of their burns infected.

When Rodney could speak, Bertoncello told him that while he'd probably never have to stand up in court, the police would need a statement. Rodney agreed to give one, but only at the same time he told his kids.

'Are you sure you want me there for that?' Bertoncello asked. He felt less an intruder than a torturer.

'I'm only saying it once,' Rodney replied.

On the day of the statement, grief was in the air like a chemical element; it was difficult to breathe or swallow. Rodney's children sat beside him, and Bertoncello turned on his tape recorder and tried to disappear into the wall.

He already knew the shape of the story:

Leatham, a carpenter, is working on his house in Morwell when he sees smoke rising, an unlikely skyscraper, over Churchill, ten kilometres away. He worries it is heading towards the tiny hill community of Callignee (pop. 367), where his daughter lives on a bush block with her partner and small children. Rodney and his wife, Annette, drive over to assist in case there are spot fires. Annette is a frail woman with an autoimmune disease. She stays inside helping her daughter, while outside her husband and son-in-law connect a generator Rodney has brought, should they lose power. Then the two men fill buckets and containers with water.

Throughout the afternoon, the family listen to the radio and check the websites of the Country Fire Authority and the Department of Sustainability and Environment. There are now blazes all around the state, although no specific warnings are issued for their area. Outside it is growing dark. Smoke blocks the sun and the sky glows red. They lose power: the lights and radio go off, the phone and internet stop working. Rodney and his son-in-law believe that if the wind changes, the long thin fire will turn like a snake and bypass their property. They've prepared for fire to come their way while believing it won't. But in many minds, staying to defend your house is the Australian test of grit: it's proof that you deserve to be living in the bush in the first place. Holding their nerve, they decide to make dinner.

In the background they can hear the blaze, constant like an ocean. Surrounded by steep gullies, they can't see flames. They can't tell where this fire is, until suddenly it feels very close. The family debate whether to stay or go, stay or go, and then it is clear they have only moments to leave.

Leatham's daughter and son-in-law drive away first in separate cars. But a beast has found them: at the end of the driveway, a spot fire ignites in the next paddock. Then, all at once, burning debris – not just airborne embers but flaming branches – falls everywhere, along with fat drops of black rain. This fire is now creating its own weather system. His daughter is driving underneath a pyrocumulus, a massive grey fire cloud that's formed over the smoke plume. Hot air has risen in a convection column, and as the cloud grows heavy it rains – pointless, ironic drops.

Black splodges of liquid soot fall over the windscreen, the wipers now cutting up the view of fire everywhere. Native animals come down the road, fleeing a burning fauna reserve. In the first car, Leatham's son-in-law hits a kangaroo, then his daughter hits it too. In the chaos, she realises her parents aren't behind her and flashes her headlights. Her partner thinks she is telling him to drive faster. At the top of a hill she stops, debating whether to return to her parents, or go on towards her children. She steers away from the fire . . .

Now, in this room full of medical equipment, Rodney is telling them what happened at the house. He jumps from past to present tense, as if he's still living each moment. His children ask no questions. They cry as he describes going to disconnect his generator, he and their mother getting into the ute, finding they're surrounded by flames – Bertoncello's tape recorder stops working and he starts transcribing Leatham's words into his notebook as fast as he can:

This is where everything turned into milliseconds. Everything is slow. Less than half a minute, a quarter of a minute and red-hot bullets are falling, cinders landing everywhere. I backed away from the house to get around a retaining wall, and there was a massive grass fire in front of me . . . The heat and wind were astronomical. I backed down to the grass and had to drive through another fast, furious grass fire. Red fire bullets were everywhere, hitting the car, falling all around.

I was trying to locate the driveway, to get back onto it. I could hardly see out of the windscreen, soot, ash and shit were all over the ute. I had to stop. I had to slow to a stop. I think the ute stalled, I don't know. It is all in milliseconds.

In a millisecond, I decided I'd back into the dam: that might be okay . . . The fire . . . the fire was coming over the hill. In the next millisecond, no sooner had the ute stopped, Annette tells me, 'Let's run to the house.' She opens the door. There's no time to say yes or no. It's just what we're doing.

She turned out of the car, out of the passenger door and fell over. I heard her cry out. I got out, ran around the car and she was virtually on fire. I tried to drag her. She was in flames. I was putting my hands in flames, but I couldn't hold on. It was so hot. I couldn't do anything. I looked around. Shrubs were like glow bombs on fire. I wanted to help her . . . there was nothing I could do.

I knew I had to move, to run to the house. The flames were head height from the ground. I don't know how I got there . . . I sat in a child's plastic sandpit shell filled with water, and cursed everything under the sun.

Bertoncello wrote this down through tears. He'd seen the aerial photographs of the crime scene. The house had somehow remained untouched. It stood there ringed with burnt earth, the ute sunk in post-nuclear ash. Nearby was the dam, from the sky a pockmark, to which Rodney had run from the plastic sandpit and submerged himself. Lying low, in a grief-filled hallucination, he'd seen the eucalypts:

glowing like Christmas trees, like somebody had put a massive amount of fairy lights over trees 30 metres tall.

I don't know what happened to the ute. I don't know if it stalled or not. All I could do was have plans . . . plan A to plan Z, and if one fails you go to the next one. Plan A – gone. Plan B – gone. Plan C – get Annette . . . She was on fire. I was on fire. Next plan . . . all of this in milliseconds.

Later, much later, the detective will handle the statements of over six hundred witnesses and slowly piece together their stories. Just after 1.30 pm on that Saturday, a group of volunteer firefighters were standing outside the Churchill fire station, smoking and yarning in the searing dry heat. In a place where there wasn't a lot to do, the building was as much a social club as a fire station, and some members had been waiting all morning, conscious they could be needed.

Everyone had heard the endless dire weather predictions ('Don't go out . . . don't travel . . . stay at home,' the premier, John Brumby, had warned.) Then, through the shimmering streets, a woman drove up in a white SUV, kids strapped in the back, and pointed to a column of black smoke rising in the hills behind them.

Just moments before, the crew had been looking at that same innocuous spot, bitching about the thought-cancelling temperature as sweat gathered under their heavy uniforms. They might have been standing in a dream – barely peopled, with those making cameo appearances moving slowly in the oven-like air – and here was the moment it turned to a nightmare, with the smoke rising ever darker, higher. This was the fire everyone had half expected. The fire they'd felt coming towards them all day.

The volunteers hurried onto the only available engine – a pumper, designed for urban firefighting – and roared towards the billowing smoke. It was the colour of a car tyre burning – the plume normally seen on a blaze that had been raging for hours, not a couple of minutes. As they reached the intersection of Glendonald Road and Jellef's Outlet, they realised there were two separate fires in the eucalypt plantation. Newborn furies, mewling and thrashing through the trees on either side of the outlet. Those on the pumper immediately had their suspicions about how these giant twins had come to life. The fire was slightly less advanced on the right-hand side, but on the left it was crowning, surging through the canopies of thirty-metre-high gum trees, propelled by the parched timber and limitless supply of eucalypt oil.

'The flames were lying down,' a crew member later told police, 'because the wind was howling through' – horizontal yellow and red flames, so fierce they looked to be rolling up the hillside. In these low mountain ranges, for every ten degrees that the slope increased, the fire doubled its speed, preheating the fuels above, causing the flames to lick faster at leaves and branches.

Even within the truck, the volunteers could feel the radiant heat coming off the blaze. None of them had ever seen a wildfire build so quickly. It was already impossible to make out the depth to which it had blasted into the plantation. They were skirting the fire's flank and they knew it was too big for any single engine to fight.

Instead they drove along the winding, semi-rural Glendonald Road, in an area known as Hazelwood North, sounding the sirens and air horns to warn people to evacuate.

The Country Fire Authority captain radioed in, asking what the situation was. A volunteer lieutenant, watching the fire approaching in the engine's side mirror, told him, 'You know that four-letter word we can't use over the radio?'

It was 'SHIT', an acronym for 'Send Help! It's Terrible'. The lieutenant called for aircraft support and twenty fire engines. Soon a CFA helicopter started dropping water from the sky.

The fire rushed south-east and its speed created a narrow fire front, but over the afternoon the flanks grew to a length of fifteen kilometres. Firefighters concentrated on these walls of flames, water-bombing them, bulldozing in firebreaks, siphoning dam water and hosing it onto the blazes left after the free-burning front had moved through.

At around 6 pm that Saturday, the wind changed direction. This, Detective Bertoncello was to learn, was the pattern with all the most devastating Australian fires: the exsiccating north-westerly collided with a south-westerly buster gusting up to seventy kilometres an hour. The cold front fed the fire a new store of oxygen as it struck the blaze on its eastern side, turning this flank into a vast front, accelerating in tandem with the mountainous terrain. Sending burning firebrands kilometres ahead, the fire front charged north-east towards the tiny townships of Koornalla, Callignee, Callignee North and Callignee South.

The Heartbreak Hills, locals once called this steep, poor country. The south-eastern corner of the Strzeleckis was opened up to European settlement at the end of the nineteenth century. Within years of a great forest coming down, many settlers had retreated from wet and sunless winters, rabbits, weeds, and pitiful roads far from any markets. Around these collapsed communities a version of the bush grew back, and now the fire, enlivened by shearing, twisting, whirling winds, swept deep into the hills towards these reduced places. Later, Detective Bertoncello would meet tree-changers who'd lived here, but more often people who had been skint for generations, and for whom, as they struggled to feed their children, an insurance policy was no priority.

Earlier, at 1.45 pm, the CFA's regional incident control centre had issued an urgent warning about an impending wind change, but there were disastrous problems with communication.

Fifty tankers were using just one fire ground channel and one command channel. These channels had jammed from being inundated with calls, or there was static from the smoke. Radios and mobile phones had poor coverage in the hill country, delaying for hours messages sent by the paging system or SMS. Vital information did not pass up or down the chain of command. Those at the top didn't know where fire engines had been deployed. Most brigades also had no idea about the timing or consequences of the wind change, and the public warnings issued to many of the communities directly in the fire's path were inadequate. People had to rely on television or radio bulletins that were hours old, not realising that an inferno was tearing their way.

This is how those in the Churchill fire's path described it to the Arson Squad detectives: 'We all heard a noise start, and it was getting louder. It was like a jet engine. I'd never heard a noise like it, and then the penny dropped – it was the fire coming'; 'We couldn't see it but we could hear a sound like continuous thunder'; 'a God-awful roar, deafening'; 'like a 747 plane sitting on a tarmac revving its engines'; 'like seven jumbos landing on the roof'; 'like seventeen freight trains'; 'this massive constant roar'; 'the noise still haunts me. It got louder and louder and it became ear-splitting.'

'It was like someone flicked a switch.' Within moments, 'the wind changed direction and it was just wild roaring'; 'it was a hurricane coming through. Trees were losing branches, and they were crashing to the ground.' Flocks of birds tore out, and wallabies and kangaroos fled the fire's path. 'As the front approached, everything started to shake . . . I was picked up off my feet and went about a metre and a half in the air.'

Soon 'you couldn't open your eyes properly, because of the smoke and ash'; 'It got dark so fucking quickly it wasn't funny'; 'daylight, dark . . . boom'; 'darker than night'; 'at one point, I was only about fifty metres from the house, putting out a spot fire, and I could not see where the house was'; 'the sky was black'; 'and then, as the main fire front neared, it started getting lighter again, the colours started changing and it went from dusky yellow to a reddish colour'; 'when the fire front actually arrived it was almost like a sunrise. The whole western sky, as far as one could see, was aflame.'

'We were in an elevated area and could see over the tree tops, the flames just coming'; 'moving very fast like someone had poured petrol on the ground'; 'Within twenty or thirty seconds everything was just exploding all around us'; 'It felt like it was raining fire'; 'red snow'; 'fine embers were blown by the wind everywhere like snowflakes'; 'little sparks falling on my skin'.

'The wind was moving in all possible directions and the embers started coming from everywhere . . . in all sizes and sometimes they were lumps of wood flying through the air: branches, leaves and smaller loose items'; 'I was being pelted with burning gum nuts'; 'It was basically hailing fire.'

'The embers became like showers of flaming arrows the size of tennis balls hurtling everywhere and bursting into flames . . .'; 'There were embers the size of dinner plates coming down'; 'embers the size of pillows'; 'as they were landing they were exploding up about seven feet in the air in flames'.

One man saw his beehives combust from the sheer heat. 'Trees ignited from the ground up in one blast, like they were self-exploding.' Burning birds fell from trees, igniting the ground where they landed. 'Everything was on fire, plants, fence posts, tree stumps, wood chip mulch, the inflatable pool. I put water on it, but it melted slowly to nothing.' The aluminium tray of a ute 'ran in rivulets on the ground'.

'It was so hot in the fire that the plastic breather in the middle of my face mask melted and the liquid plastic burnt my lips. I grabbed my sunglasses, and they actually squashed and melted in my hands.' That night, this man slept upright to avoid the painful weight of his eyelids.

Inside houses, people took shelter while '[t]he heat coming through the front windows was amazing'; 'I felt like I was inside a Coonara [a wood heater] looking out from within the fire'; 'all I could see was red with little black sticks flying through the air'; 'everything was blood red and you couldn't get a depth perception of how far this blood red fell'; 'It was like the air was red. There was no air in the air'; 'It was like sucking on a hairdryer if you tried to take a breath'; 'you could feel your skin melting from the heat of it'. From their houses, people watched fireballs coming towards them. One man went out into the inferno with a gun and shot his horses.

Another man admitted: 'When I saw the fire, I was initially mesmerised by the sight of it. Our house had floor to ceiling windows and I had a full view of the flames, which, at a guess, would have been at least thirty metres high, moving horizontally, smacking into the side of the house and wrapping around it. It was as though the house had been picked up and thrown into a sea of fire . . . I noticed at about that time that our water pump had failed and it was then I realised we were in "Scenario Z" – or in other words, that we should not have been there.'

Windows started to crack, then curtains were ablaze; skylights melted and began to drip; fire came under doors, or through the subfloor or the ceiling. The bins and buckets and plastic containers that people had filled with water seemed absurd. One man went outside to check if it was safe and came back with his boots on fire. Someone else ran from the flames and realised his jeans had ignited. People stayed alive by breathing through wet fabric, or lying in dams or creek beds. A man waited in a fish pond with a tea towel over his head.

Another man and his son survived in a dam by 'grabbing lily pads and putting them all around our faces and over our heads, armfuls of lily pads, even the green slime helped': any hair or skin uncovered was singed. Kangaroos joined them in the water as the main front passed. 'At one stage,' this same man said, 'I looked up and I could see a blanket of flame that went from one tree line to the next. It was like someone was waving an orange blanket over our heads . . . you could almost touch it.'

A brigade of firefighters caught in the eye of the firestorm scrambled into their truck and held a fire-retardant covering over themselves while two of them set their hoses to 'fog spray', creating a secondary layer of mist. One man holding a hose tried to breathe through his wet gloves. When he dared glimpse outside, 'it was like looking into a furnace', a hell, 'hell would be better actually'. They heard the low-water alarm.

'Mayday. Mayday. Mayday!' the man in the driver's seat called on the radio. 'Full entrapment and under fierce attack.'

'Understood,' a voice came back. 'There is nothing we can do for you.'

'We understand.'

Someone asked, 'Anyone go to church?'

'Not fucken likely.'

They drove to the door of the house they'd just been defending and ran inside to collapse on the floor, weeping. None of them could understand how this place was still standing. But the fire – fickle, picky – left them to live.

The din of flame-engulfed buildings was a noise one woman later heard repeatedly in nightmares: 'it sounded like they were screaming. It was so loud.'

Another group of firefighters broke into a house with an indoor pool and survived in the water with the pool's rubber heating hoses alight and burning next to them, the house's fire alarm blaring like a sick joke.

Two brothers, Colin Gibson, forty-nine, and David, eighteen months younger, had both been volunteers in country fire brigades. They'd met that afternoon at their elderly parents' log cabin on Glendonald Road, a kilometre up the street from where the fire had been lit, to prepare its defence. They blocked off the house's downpipes, filled the gutters with water, hosed down the walls, and left water-filled garbage bins and wet cushions on the verandah. Their family called continually to monitor the situation.

And then the phone calls went unanswered. In the middle of the night, David's daughter avoided the police barricade by driving through the plantation's service tracks. She veered up her grandparents' driveway and assumed, in the dark, she'd taken a wrong turn. The house was missing, turned to rubble.

As the fire raced towards his farm in the steep hills of Jeeralang North, 77-year-old Erich Martin filled a wheelbarrow with precious possessions, which he was soon pushing outside as he and his wife, Trudi, eighty, fled their burning house. Then Erich noticed flames licking at the wheelbarrow. He moved it to their orchard, which was already blackened, and came back to find his wife on the ground, stretched out, he later told the police, as if 'lying on the beach'. She looked so peaceful that at first he thought she was resting. She'd had a heart attack and died.

Eight kilometres east of the fire's beginning, in Koornalla, Alan and Miros Jacobs, fifty-two and fifty, had spent the afternoon preparing to defend their house from fire. Alongside them was Luke, their 22-year-old son, who phoned a friend to ask for extra help. The friend, who was at an eighteenth birthday party recruited a group of others, the call to arms punctuating this endless summer day. Between bouts of fire preparation, the young friends cooled off in a swimming pool. Out of it, the air was scorching.

The Jacobs owned an equipment supplies business, and there was a forklift in the driveway. One friend lifted another up to see if he could spot flames, but there was just smoke, great clouds of it. When embers started drifting in, most of the helpers decided to leave. Then it was just the Jacobs and 21-year-old Nathan Charles, a part-time scaffolder, who felt it was right to stay. They fought the fire that soon arrived for as long as possible before seeking shelter in a homemade bunker under their house.

Around 6.30 pm, Charles phoned his father, a truck driver just returning to the Valley from an interstate job, who thought Charles sounded like he was saying goodbye. The call dropped out. The father dialled 000, waited on hold for an eternity, then drove to the Hazelwood North fire station and begged the CFA members on duty to help his son. They told him to ring a central number. He felt he would collapse and die himself right there. He rang his partner and said, 'I think I'm about to bury my son.'

A text message soon arrived:

Dad im dead I love u

Further east, on Old Callignee Road, were another father and son. Alfred and Scott Frendo, fifty-eight and twenty-seven years old, gave up on the family home they had been trying to defend and fled in their cars. Their two vehicles were later found sitting on burnt steel rims, moored in ash, one and a half kilometres from their house, which remained undamaged.

The car of Martin Schultz, thirty-three, was found the following week by a Callignee farmer who was dragging his dead cattle along the scorched ground to a burial pit. The farmer had lost his house, animals, sheds, fences, pasture. All he'd saved were three drawers of photographs. He saw the steel frame of the car sticking out of a creek bed. Molten silver metal had flowed from the vehicle's chassis and solidified on the ground. Schultz, who worked in a local brick factory, had been fleeing with his own photos, those of his son as a baby. He called his father-in-law, who was minding the boy, to say his car was on fire. Then the call dropped out.

In Callignee, fifteen kilometres from the inferno's beginning, a man and a woman, her daughter and her daughter's boyfriend survived the fire front by lying on the floor of their shaking house with wet towels over their faces. Between the waves of flames, the man had gone outside to extinguish spot fires around the house. During one of these sessions, a figure came towards him.

Thinking it was his neighbour, he called, 'Good on you! You made it through.'

It was the neighbour's father-in-law. 'I've lost my wife and I'm burnt!'

The first man rushed and grabbed him, helping him onto the verandah. In the dark, the daughter could just make out the visitor's smouldering clothes and she smelled burnt flesh. He told them his trousers were on fire. In fact he was wet from sheltering in the dam, but they took them off him anyway. He couldn't feel his blistering legs.

The family put wet towels on a deckchair and helped him sit down. They filled up a wheelbarrow with water and put his feet in it, wrapped in more towels. The daughter dipped sponges into the wheelbarrow and wrung the liquid over his whole body.

'My wife,' he kept crying, 'my wife!'

He told them he'd grasped her hand but wasn't strong enough to hold on. It felt as though he were standing inside a volcano. The flames scorching him, his skin dripping from his body, he had to let her go.

Both the man's hands were now badly burnt. One little finger had turned to jelly; it was translucent when he held it to the light. Carefully, the daughter took off his wedding band before the ring finger swelled further.

He told them how long he had been married and said, 'Why couldn't it have been me? I just want to hold her one more time.'

The ambulance wouldn't come. It was too dangerous. Burning trees lay across the roads, houses were still alight.

The man began hallucinating. Every time he was given a drink of water he would tell his hosts to give his wife a drink as well. Soon he thought this family were members of his own. He glanced at nearby flare-ups, telling the flames to fuck off and leave them alone. The fire he was seeing was a creature, a demon. It was alive.

He began to scream with pain.

Three hours after his arrival, two CFA volunteers broke with orders and drove the hazardous roads that wound to the house.

Outside it was pitch-black, but trees and sheds were still burning and in the firelight they made out an ash-covered car – keys in the ignition for the planned escape – connected to a horse float, the horses now dead in the stables.

They walked through a burnt garden, and on the verandah, under a melted laser light, a man sat shaking, draped in wet towels from head to toe. Using a fire blanket, they lifted and carried him to the back seat of their ute. They drove him away along roads walled by flames, passing the shells of burnt-out vehicles. A skeleton in one sat upright in the driver's seat.

An ambulance was waiting for them at the Traralgon South fire station. The paramedics asked the man about his pain level and he answered, 'A hundred out of ten.' Later, he believed that he'd actually felt nothing, that the physical pain didn't matter. They gave him morphine and the next thing he knew he'd woken a fortnight later in an orange hospital room.

By this point, the Arson Squad had arrested someone for lighting the fire but the news was difficult for Rodney Leatham to take in. He was still coming to after days and nights of strange dreams. He'd emerged from his coma knowing which part of these dreams was real. 'The easiest thing in the world would have been to cuddle my wife . . .' Detective Bertoncello heard him telling their children. 'But no – I had to be stronger and I was.'

The Arson Squad had set up an office in the Morwell police station. This new grey modernist building, larger than might be expected for the region's population, was just off the main street. The chimney towers of Hazelwood were visible from the windows, as they were from nearly everywhere in the coal town. In the old days, the power station's bell rang at 7.30 am, then at 4.30 pm, and the men trudged to and from work. Most of those jobs were gone now. These days the station sat in the midst of a flaring crisis at the best of times, and, after the fire, it was surrounded by an alien landscape. Even when the brand-new toilets were flushed, the water was black. Local officers who were drilled in the day-in, day-out crimes of family violence and abuse now met with the newly homeless and the relatives of the missing, anguish not yet settled on their faces.

It's basic policing theory that if you follow a chain and it yields no answers you return to the start. Four days after the fire had begun, on Wednesday 11 February, Detective Adam Henry decided to revisit the area of the blaze's origin, half hoping some pointer might be lying there unnoticed. The day before, he had taken a police helicopter over the Valley and seen where the fire had travelled. It was a form of vertigo having the ground come up to meet him, for the day before that, on the Monday, along with the forensic team, he'd visited the places people died. He saw close up the torturous things the fire had done. Now he wanted to look again around the plantation.

Paul Bertoncello accompanied him, and in the unmarked Land Cruiser they drove back towards Churchill, passing a blur of gas turbines, coal-handling buildings, electricity wires and towers. Closer to the township, a sign pointed to the campus of Monash University; there were a few shops and a supermarket. Not a lot to see, only the uniform houses, coated grey from ash and the lingering smoke.

At the plantation they met Ross Pridgeon, the DSE fire investigator. Henry led them behind the crime scene tape, amongst the scorched eucalypts. The chimney reek was inescapable. The men walked around in the blackness, taking in the two marked areas of ignition. The bright little flags like bunting, pinned in the soot.

It was the first time Bertoncello had left the Morwell station, other than to sleep, in nearly seventy-two hours. While Henry had driven straight to this area last Sunday, Bertoncello had been directed to get an incident room ready for the operation at the station. It was hard not to be overwhelmed by the scale of the devastation. Soon it would be known as Black Saturday: four hundred separate fires had burned in Victoria, giving off the equivalent of 80 000 kilowatts of heat, or 500 atomic bombs. Bertoncello quickly began to learn every acronym of emergency management terminology. Contacting the local ICC (Incident Control Centre), he met the EMT (Emergency Management Team) and a dozen other sets of initials, and tried to work out who was doing what so that the Arson Squad could concentrate solely on the criminal investigation.

Now he walked among the charred trees, looking at the burnt ground, trying to think, just think. There was only one road in and one road out. The first and obvious line of inquiry was to locate witnesses, to doorknock in the area. Those living nearby might have information about unfamiliar people or cars being around on Saturday, or perhaps the first firefighting crew had seen something. Bertoncello was already plotting out the next move.

He was a tall, slim man in his early thirties; his perfect baldness accentuated his facial features and rightly gave him a cerebral cast. In his spare time, Bertoncello chose to do jigsaws, Sudoku, and other logic puzzles. Sometimes he might stare at one for two or three days – a problem as inexplicable as this scorched scene – and not write a thing down, then at some point it would click. He'd piece together the right bits. It would all make sense. He was prevented now, though, from seeing the vast complexity

of the damage by the surrounding hills: the topography compart-
mentalised the view of the destruction.

Adam Henry had sat beside the crime scene photographer
as the helicopter flew over this patch of ground. Hovered above
the areas of origin, Henry saw the two deepening V-shapes
where the fires had started, and then ash to the horizon. Oddly
sensuous shapes unfurled underneath him as they flew low over
gullies and crevices and rises, revealing houses that were fire-flat-
tened, dead wildlife and farm animals, the shells of tractors and
farm equipment, burnt fences, and surviving livestock wandering
the debris-stricken roads, eating any vestiges of green.

The detective directed the photographer to each rectangle of
burnt land that needed documenting. This was a version of omni-
presence, seeing the death scenes he had investigated twenty-four
hours earlier, but with no godlike power to intervene.

From the air, some houses looked to have been peeled. A roof
skinned off one revealed a blueprint of ash. The spaces in which
a family had slept, ate, washed were demarcated in black and
white. From one angle, the rooms may have been the chambers
of the heart. It was a mental exercise to see all this horror and not
keep asking, almost as a tic: *Why? Who?*

For all the science, Adam Henry knew that arson was a crime
of which the Arson Squad – like everyone else – knew very little.
In the mid-nineteenth century, pyromania was considered to be
'a morbid propensity to incendiarism, where the mind, though
otherwise sound is borne on by an invisible power to the commission of this crime that is now generally recognized as a distinct
form of insanity'.

Over the 75-year history of the mental health bible, the *DSM –
Diagnostic and Statistical Manual of Mental Disorders* – the
classification of pyromania has fallen in and out of fashion and
different editions' pages. Today, of the multitude of people who
deliberately light fires, only a scarce few are considered to have
an all-consuming 'fascination in, curiosity about and attrac-
tion to fire' teamed with 'pleasure and relief when setting fires'.
The behaviour is now believed to be better accounted for by the
DSM's section on disruptive impulse control and conduct disorders. An individual's tilt towards the antisocial and a mad lack
of restraint.

Through the years, various agencies have tried to establish
criteria for profiling fire-setters. But most international studies
focus on the deliberate ignition of houses, cars and buildings,
rather than wildfire arson, a form of fire-lighting that, although
not unique to Australia, is a national specialty. Of vegetation
fires in this country, 37 per cent are deemed suspicious, and
13 per cent maliciously lit (whereas 35 per cent are considered
accidental, 5 per cent due to natural causes, and another 5 per
cent due to reignition or spot fires. The rest are shelved under
'other causes'.)

Adam Henry knew the basic hypotheses of the FBI and
various other profiling systems, and was conscious some were
fairly complicated. One prominent model used this equation
to explain the behaviour: FIRE-SETTING = GI + G2 + E, WHERE
$[E = C + CF + DI + D2 + D3 + FI + F2 + F3 + REX + RIN]$.

What the sum tended to find was that fire-setters were more
often than not male; they were commonly unemployed, or had a

complicated work history; they were likely to have disadvantaged social backgrounds, often with a family history of pathology, addiction and physical abuse; and many exhibited poor social or interpersonal skills. It was a plausible profile, but hardly different from that of many non-firesetting criminals. In other words, close to useless.

The Arson Squad was aware that there were more deliberately lit fires near the urban–rural fringe – places where high youth unemployment, child abuse and neglect, intergenerational welfare dependency and poor public transport met the margins of the bush, the eucalypts. And that pretty much described most of the towns in the Latrobe Valley.

Living back in Morwell, you were three times more likely than the state average to experience long-term unemployment (22 per cent of kids lived in jobless families). The rate of kids in out-of-home care was the highest in Victoria, as was the rate of crimes committed with children present. You were 2.6 times more likely to be a victim of domestic violence, and the rate of substantiated child abuse was three times the average. All these statistics that hid as much as they revealed.

Three days after Henry's helicopter ride, local detectives would arrest a man they found dirty, with matted hair, on eighty-three incest offences. DNA testing proved he'd fathered four children with his daughter.

The area over which Henry had flown was full of what appeared to have been relatively peaceful bush blocks. That's what it was like here: God's own country alongside that of the beleaguered. As the helicopter turned, the trees became charred diagonals, and with each undulation the hills held up their damage for inspection.

Henry had grown up in the country himself. He knew you could feel a stirring, alone out in the bush. With no witnesses to your good deeds or your bad, the isolation gave you licence. Along a lonely bush road, what was to stop a person dropping a lit match, or leaning out the car window with a barbecue lighter? Henry looked upon the blackness beneath him and thought the arsonist could have been anyone.

From the air, the detective had noticed something strange about the pattern of the burnt land near where the fires had been lit. Although the blaze had initially stretched south-east from the eucalypt plantation, in the helicopter he realised there were also extensive scorched areas to the north of the ignition sites, behind those pristine *radiata* pines. He'd wondered if there could have been a third deliberately lit fire, whether the arsonist, after setting the first two, had kept going.

Now he, Bertoncello and Pridgeon were testing the theory. They drove along an access track through the pine plantation and parked near a dry creek bed. From here they walked towards the area that Henry had observed. Smoke was still rising from these trees, both standing and fallen. The undergrowth of scrub, wire-grass and blackberry was blackened or devoid of natural colour; leaves were crisp and brittle to the touch. The men started search-ing again for signs of the head fire, seeing if they could figure out the path the arsonist had taken.

Pridgeon wrote in his notebook: 'can't be too many ignition points as fire behaviour would be extreme. ie, need to be out of area quickly'. Saturday had been a day of such heightened alert, the arsonist must have known that plantation workers would be doing patrols, and that the CFA would soon be coming. Also, on the edge of the eucalypt forest there would have been an overwhelming scent of the trees' oil in the air. If, in that explosive atmosphere, the odour was petrol you would run. Perhaps that was part of the thrill, but with the fire itself so dangerous, lighting multiple blazes could have left the firebug entrapped.

Standing amidst the burnt pine trees, Bertoncello had noticed a farmhouse on the top of a hill to the east. The house appeared undamaged, as if it had landed on a black planet. While Henry and Pridgeon kept searching for signs of a flanking or head fire, he decided to visit the house's owners. He wanted to ask whether they'd been home around 1.30 pm on Saturday and if the area under suspicion had blazed then, or if the fire had doubled back here with the 6 pm wind change.

Walking up the steep hill, he became aware of a different kind of smell. Not burnt timber, but something acrid. The odour of incinerated building materials. He'd smelled it before as a young constable, turning up to house fires, and he'd done a lot more of that in his six months with the Arson Squad.

The previous week, Bertonello had been working in Morwell, assisting Taskforce Ignis with the investigation of the recent Delburn arson – the series of eight deliberately lit fires, three of which had joined up and burnt out of control. That had become a massive fire: until Saturday, the state's worst in years. A $100 000 reward, big money around here, was being offered for information leading to an arrest. The Taskforce Ignis detectives had established that a suspected arsonist was in the vicinity of six of the Delburn fires. They knew this man wasn't responsible for the Churchill blaze because he'd been under surveillance on Saturday. That was one suspect ruled out, but at the Morwell police station the phone had been ringing constantly with locals calling in their tips on other possible candidates.

Victoria Police had no centralised database containing known or suspected fire-setters. Regional police kept their own spreadsheets on locals they were worried about, but this information was often ad hoc, disjointed. As part of the Delburn case, Bertoncello had been sifting through the names of more than thirty people believed to have previously lit fires throughout the Valley.

In the following week, he was to come upon a small item in the newspapers about a woman on the list. Gippsland firefighters had converged on Morwell to protest the three-year jail sentence given to Rosemary Harris, who, two summers earlier, in nearby Driffield, was caught lighting fires with her son in the bush. The firefighters – some still waking in the night, crying, after Black Saturday – believed she deserved longer. At the time of the fires, she was twenty-nine and the boy was fifteen. Her other six children had been waiting in the car. At the sentencing, she was pregnant with her eighth child, and the rest had all been removed from her care.

Bertoncello reached the top of the steep hill and approached the farmhouse: weatherboard, raised on stumps. Just a few doors up the road, the Gibson brothers had died defending their parents' house. Bertoncello assumed that in a community like this the residents knew their neighbours, and would be in shock. It was just before 11 am when he knocked on the door.

Liam Ferguson, a student in his early twenties, answered. He was still adjusting to the landscape he now saw when opening the door. Every window also had a new view.

He told Bertoncello that he had been at home – home as it was once known – on Saturday around 1.30 pm, with his parents and sisters. Suddenly they had heard the helicopters water-bombing the fire barely a kilometre away. His mother and sisters evacuated. Liam; his father, Tony; his brother-in-law, Peter Moretti; and Peter's father, Ray, had stayed and worked through the blistering afternoon, wetting the house down, filling containers with water and blocking up drainpipes, in case the fire came their way. And it did: the 6 pm wind change brought the fire right to the house. The men had done enough to save the building, but everything else was burned black.

So it *was* the wind change. There was no third deliberately lit blaze. Bertoncello thanked Liam and started to retreat, but the young man had more to say.

He had been meaning to come down to the police station, he admitted, because that evening, alongside the terror of the fire, something strange had occurred.

It happened around the time the wind turned. At about 6.20 pm the forceful northerly that had gusted all afternoon suddenly stopped and the sky was still, silent. Liam, who was defending an area of the property by himself, waited. Minutes seemed to pass. A wall of new wind, 'very, very strong and very, very loud', came pushing from the south, steering the fire directly towards the farm. Smoke blocked the sun. Embers flew into the dark. Liam had stood with a puny garden hose in hand as spot fires ignited all around him.

Through the haze, he saw a silhouette walking from the driveway. At first, with relief, he thought it was his father. 'We need help!' Liam screamed, but the wind's howling meant he could barely hear himself. The figure continued towards him, past the flames of the spot fires, morphing into a stranger: a man in his late thirties with a stocky build and a pudgy, boyish face. He was dressed in camouflage-print clothing – T-shirt, shorts, hat – and heavy workboots. He was carrying a dog.

Liam assumed the man was trying to escape the fire, only to find more of it. But at that moment, his hose ran out of water pressure. He needed to find shelter quickly and sprinted through the blinding smoke, not realising the stranger was following him until they'd both made it safely inside the house. This man stood quite calmly in the living room, cradling the small, tan-coloured terrier as if it were a baby, its soft stomach in the air.

As the fire roared next to them he told Liam, in a voice very slow and distinctive, that his car had just broken down nearby. Right then Liam was too panicked to care. He threw the man a phone and ordered him to call 000, while he went to search for his father. Surrounded by flames, it was hard to see, or breathe, or imagine survival.

Ray Moretti later told police that he had run back to the house to seek shelter just after Liam left. A fireball was rolling towards them and he felt increasingly desperate. On the back verandah he found someone he did not know holding a dog. 'I basically said, "Get the fuck inside!"' Ray slammed the door behind them, jamming it on the hose he carried. When it seemed the fire front had passed and it was safe enough to leave, they ventured back out. The pump was working now, and Ray directed the man to help feed him the hose. They needed to stop it getting caught in rose bushes. The stranger tied up his dog and did as Ray told him.

His name was Brendan, he said. His car had stopped just down the road. He couldn't afford a new car and was worried the insurers would only pay five hundred dollars if it burned.

Ray thought this Brendan had 'a dull look'. He seemed vague and was silent unless spoken to. But down and up the hill, through the garden's rose bushes, he followed Ray as embers rained upon them, keeping the hose untangled to put out spot fires – in truth, he was of great assistance.

When finally Liam found his father, and there was a moment to exchange more than a few words about anything other than the fire, he told him about their visitor. It was odd, freakish even, to turn up in the middle of a firestorm, and Tony went to check the stranger out.

Tony subsequently told police that he'd walked through the house calling, 'Hello, is there anyone there?' Outside, down from the verandah, stood Brendan. He said he was a friend of their neighbour, Peter Townsend; they'd once been gardeners together at the local university. Tony Ferguson was pleased to have another set of hands helping to defend the property, but even in this surreal moment – his house still standing in a newly scorched world – his guest's presence was incongruous.

The house was over the back of a rise, so Brendan had had to walk up a random gravel road while helicopters water-bombed around him; there was burnt bush on one side, flames coming up behind him and potentially up in front too. He had passed the CFA firefighters, who felt it was too dangerous to approach the property.

'Shit, you're game!' Tony exclaimed when Brendan told him his story. 'I wouldn't do that.' The route had been practically suicidal, and later Ferguson wondered if that had been on the man's mind, the very reason he'd walked in.

Around 1 am, after a five hour ember-storm, a CFA fire truck finally ventured up the drive. Tony Ferguson asked the crew if they'd give Brendan a lift home. This man in his camouflage gear made him uneasy. 'We had enough to deal with,' he told police, 'and we had to count on everyone who was with us, so I felt uncomfortable.'

Liam Ferguson now handed Detective Bertoncello a plastic bag containing a camouflage-print canvas hat. Brendan had used it as a water bowl for his dog, then left it behind.

By this point Adam Henry and Ross Pridgeon had arrived at the farmhouse by car. They learnt there hadn't been a third ignition point, and Bertoncello gave Liam his card.

Pridgeon peeled off and the detectives went to check the crime scene at the Gibson property up the road. Surrounded by a forest of tall black sticks, they found the rubble where the house had been. The chimney still standing, an affront.

Bertoncello's phone rang. It was Liam Ferguson. He'd remembered something else. On Saturday, a sky-blue sedan had been parked at an odd angle by the grass verge of Glendonald Road. The car looked to have stopped suddenly. Liam's mother had noticed it too when she'd evacuated the area around 2 pm. After the wind change, the car had been incinerated. It was towed away the next day.

The detectives left the Gibson property to meet Liam down the road, where he thought the car had been located. A few burnt metal scraps lay nearby on the blackened ground. If these were from the car of the Fergusons' visitor, it had evidently broken down four hours before he'd come calling on them. He could therefore be placed near the area of the fire's origin within a half-hour of it starting.

The detectives were practised at looking noncommittal, but they and Liam knew what this was about. Word was already out that the fire had been deliberately lit. Liam's younger sister had seen two separate fires burning just before she evacuated with their mother. Henry and Bertoncello sensed the shock in the air shifting to fury. They looked at the car scraps and tried not to appear overly interested.

When Liam returned home, the pair began canvassing the Glendonald Road residents. These Hazelwood South bush blocks had offered nature and privacy, but now the men were walking past burnt cubby houses and trampolines, up burnt paths to houses with smoke-blackened walls and windows. They knocked on doors to check if anyone else recalled seeing the blue car.

Few residents were home. The road was still blocked off. Locals couldn't get in to survey the damage to their properties, or, if they had stayed, out to get food. Those who'd stayed half wondered where they were: all the familiar markers – trees and houses – had been torched. These people were coughing up black gunk and had burnt skin and eyes. Their children, entrusted to friends and relatives, did not want to return (or, later, to stay on hot days). It didn't feel like home anymore.

Those who answered their doors told the detectives they had noticed the blue car. It was a Holden sedan with magnum alloy wheels, parked at a strange angle, looking abandoned.

Geoffrey Wright, a sheet metal worker at the nearby gas plant, would later tell police that he had been struggling to catch his wife's horse when he heard the distinctive drone of the blue car's engine. He had noticed the sound before, when he saw the car driving up and down the street. It had gone past so often that he thought its owner was a neighbour. The drone ceased and Wright realised that the blue car was parked a little way off, as if it had just broken down.

'I saw the driver running towards his car. He got down on his hands and knees and had a look at something under his vehicle. He was in a hurry. He got up and ran around to the driver's side and tried to start the car. I felt sorry for the poor bugger. Tell you now, if I wasn't chasing the horse I was going to offer to give him a tow or a lift. I thought he was a neighbour and he was in a panic trying to get to his family.'

When the wind changed that evening, Wright had still not managed to corral the troublesome horse. Finally, he tried to drag it into the house, but his arms were burning from the radiant heat and he had to let it go. The fire brigade found the animal's carcass, but with the roads still blocked it would be another week before it could be buried.

Most of the residents had a story like that about a pet or live-stock. And so, they told the detectives, they were staggered when the very next day – while trees and fences were still burning all around and the CFA was putting out a fire two doors down – an orange tow truck arrived to take away the blue vehicle's shell. Out of the truck stepped the driver and the car's owner, as blithe as could be.

One local got into a fight with the owner. 'You can't move this, mate,' he recalled saying. 'It's part of the crime scene. It was here when the fire started.'

'Don't start that shit,' the owner replied. 'No, it wasn't. I come down to help a mate with the fires.'

The detectives left Glendonald Road, and at the police barricade stopped to ask the traffic cop which local company used orange tow trucks. It was Connolly's, back in Morwell.

On Monash Way they passed the thirty-metre golden cigar. And further along, on the strip of land between Churchill and Morwell, an inscrutable network of wires, insulators, transform-ers. Past the switching yard were more unfathomable complexes surrounded by high fences – all these countless parts assembled in the service of power, heat and light – and beyond, the smokestacks of Hazelwood Power Station, ever-present, standing guard.

They had to get back to the police station. The officer in charge of the Arson Squad, Adam Shoesmith, was about to arrive in

Morwell and needed to be briefed. But first they parked beside the corrugated-iron fence of Connolly's Towing and Panel Beating. Inside the shop, on the concrete floor, car bodies everywhere.

People feel the atmosphere change with the approach of detectives. These two were sunburnt from the morning's outing, and wore navy arson squad overalls that stank of smoke. A receptionist saw them and hurried to get her boss.

Andy Connolly put down his lunch and came out of his office. Henry and Bertoncello, not wanting to raise suspicions, told him they were trying to account for missing people. They asked if any of his trucks had recovered cars from the fire zone. Connolly said his driver had only picked up one vehicle.

He took the detectives to a sedan that looked to have been built out of rust. There was not a trace of blue paint. No upholstery, no windows, no tyres, just steel rims. It was a burnt-out 1974 HJ Holden sedan and they could barely make out the indented digits of the number plate: SLW 387. Connolly showed them a copy of the towing receipt. It was made out to a man named Brendan Sokaluk of 11 Sheoke Grove, Churchill.

Out the train's window, the views were sepia-toned. Smoke gave the unspooling landscape an old-fashioned brown hue, the filter of a dream sequence, although Detective Senior Sergeant Adam Shoesmith had barely slept in the past three days. As officer in charge of the Arson Squad, he'd coordinated the response to all the Black Saturday fires deemed suspicious.

At first, units of detectives, forensic scientists and crime scene experts were deployed to different regions: the standard response. But the devastation was on a scale that could be envisaged only by those of his colleagues with training in counter-terrorism. Blazes, believed to have been ignited by failing powerlines and arson, had burnt 450 000 hectares. Around the state there were areas that still couldn't be reached, and in areas that could, corpses were everywhere. The army and the state emergency services were finding the bodies of those who had sought shelter to no avail by the sides of roads, and under bits of tin that were formerly houses.

Detective Shoesmith had opted to ride the train to Morwell that Wednesday because he didn't want to drain resources unnecessarily. He had just sent teams of officers to various locations, accompanying the Disaster Victim Identification units. They'd need dozens of cars to drive to the families of the dead, and gather ante-mortem evidence, such as toothbrushes for DNA.

It would take weeks to establish that altogether 173 people had been killed. Eleven of the dead were in the Latrobe Valley, where Shoesmith was now travelling to take over the supervision of Operation Winston.

Other passengers stared out the train's windows at the scorched ground with what looked like numb incredulity. People had understood that this *could* happen, but had never really believed it would. In some ways it was a shock to feel such shock.

Australia had always been a burning continent. Shoesmith, a clean-cut history buff in his late thirties, knew that. In the area through which he was travelling, the once lush Gondwanan landscape had given way over millennia to the hard-leafed eucalypts, sclerophylls dependent upon burning for propagation. A perfect loop of fire creating more fire, a dynamic that later intensified after the arrival of the first Australians.

Aborigines used fire for illumination, for signalling, for creating tracks for travel, for inducing green growth for the animals they hunted, for pursuing animals fleeing a blaze's flank, for harvesting such food as sweet tuberous yam roots, and for preventing larger conflagrations, which they could not control. Fire was used in ceremonies and there were ceremonies about fire. Flames, and their endless complex interplay with the season and flora and wind, were inseparable from daily life and culture.

Later, Aborigines harried European explorers with fire, and tried to burn out the white settlers who followed. One theory has it that the 1851 Black Thursday blaze, which scorched 5 million hectares, a quarter of the state, was a form of indigenous warfare. Later again, a glass bottle, put in the right place on a hot day, was said to have been used against a farmer who wouldn't let an Aboriginal man onto a traditional burial site.

Itinerant white shearers and swagmen would also threaten stingy pastoralists with fire, and disputes between settlers often contained the threat and frequently the deed. Postcolonial history is full of stories about flames being used less out of madness than revenge.

Europeans saw the advantages of ignition but lacked the indigenous understanding of it. Fire made 'wild', 'primeval' forest into productive farmland. The hills around the Latrobe Valley were dominated by mountain ash, towering eucalypts sixty to ninety metres high, the hollows of which were large enough to be used for church services, and, ironically, as dwellings by those who'd lost their homes in bushfires. These forests were cleared by man-made infernos. Working up the hillside, settlers scarfed the bigger trees that pushed their way through the dense understorey, and at the hilltop felled the giants, which in turn felled everything below. Thus the logs were set. Then, on the most ferociously fire-friendly day, when a hot wind blew from the north, these pioneers set a match to the lot. It was like an atomic bomb, or the end of days. And when it was over they stood knee-deep in ash. In some places they saw the sky for the first time, and discovered they had neighbours.

Detective Shoesmith had grown up in Officer, on the other side of these now treeless hills, back towards Melbourne. In those days it was still a country town rather than an outer suburb, and the netballers married the footballers and life had an easy symmetry. He had planned to study archaeology, but his application for a police cadetship succeeded a few weeks before the offer of a university place, and so now he watched the History Channel obsessively in the evenings and spent his days investigating the artefacts and biofacts of crime.

As he stepped onto the platform of Morwell station, his head was pounding with fatigue and the worry of what he'd find here.

Two Aboriginal men approached him for a cigarette. He told them he didn't smoke and they instead asked a passing white skinhead with a giant collared dog.

'Fuck off!' the skinhead replied, and a fight broke out.

Shoesmith, who'd cut his teeth in major crime and spent years knocking over meth labs, telephoned the police station, asking to be rescued.

The station was just over the road. Shoesmith was soon briefed by the most senior officer on the case, Inspector Ken Ashworth, before taking control of the operation. Shortly he was overseeing the work of dozens of investigators. His was the directing role, but he reported to Ashworth who himself reported back to Crime Command, the team that effectively ran the Victoria Police response to Black Saturday.

The Brendan Sokaluk line of inquiry sounded promising, but Shoesmith had long ago learnt to resist the urge to drop everything and chase down the sexiest lead. And realistically, Sokaluk's car could only be traced to the fire's point of origin half an hour after its ignition. The man could have been just another pyrophile out for a drive, one of the many people who'd turned up to see the fire take off. Later, the investigators noticed that a high number of local households possessed scanners, possibly, they theorised, to keep abreast of emergency services dealing with their neighbours' dramas, although the devices also alerted people quickly to dangerous blazes.

Meanwhile, the list of suspects was multiplying, some candidates more bizarre than others. One person had rung with the name of a young local logger, who the previous year had taken his seventeen-year-old girlfriend on a picnic, then staged their abduction. Face hidden in a balaclava, he cut off her clothes and bound and gagged her, before reappearing as himself to 'save' her. The two of them walked naked in circles through the bush for a week to evade their non-existent captors. (In 'brown face', with a fake Indian passport, this blond man then fled to Delhi, only to be sent straight back by unimpressed custom officials.)

Both were members of Morwell's Apostolic Church, which – not for nothing – preaches the utter depravity of human nature.

In the afternoon, while Adam Henry took Shoesmith and Ashworth out to the crime scenes, Bertoncello did some work to see where Brendan Sokaluk's path led.

This 'person of interest' had no criminal record, but it turned out that in the past he'd been the subject of intelligence reports. Bertoncello pulled them up. Two years earlier someone from the CFA had notified police that when Sokaluk tried to join up as a volunteer his demeanour was strange. Another report suggested that fires had been previously lit on Glendonald Road using teepee-style configurations of leaves and twigs. On one such occasion, Sokaluk's blue car was seen driving past the next day: had he been checking out his handiwork?

An investigator was sent to speak to Peter Townsend, whom Brendan had told the Fergusons he was visiting. Townsend, in his early fifties, was a no-nonsense farmer still living in the Victorian cottage where he'd been born. Here, in the rolling green hills of arcadian Gippsland, he tended an orchard with forty different varieties of heritage apples: an idyll that showed again the sharp edges of parallel lives. The cottage, with its symmetrical facade and stained-glass sidelights, looked drawn from a picture book, albeit one with minimal colour. All the house's weatherboards were dark with smoke damage. The orchard, near a creek bed, still stood while the hills around were blackened. Townsend had thought more than once that Saturday that he would die too, and he seemed to the investigator half stunned at his own survival.

Townsend had a crucifix by the kitchen sink and a portrait of his younger self in military uniform over the fireplace. His dark-brown moustache was unchanged from those days, and he had the upright, boxy walk of an ex-soldier, or perhaps it was the stiff gait of an apple farmer. He was not surprised to see the police.

The fire had started less than a kilometre from his cottage. Townsend had walked outside and seen flames twice as high as the tallest gum trees on the hills, west of his house. He and his wife had sprinted to reach their car, and as they were evacuating he recognised Brendan Sokaluk's Holden in the chaos. It was parked at a strange angle. What was Brendan doing out, he wondered, on a 47-degree day? Townsend had his suspicions.

All the investigators had been briefed not to appear too interested when Sokaluk's name came up. The detectives didn't yet know his network of relationships, and didn't want one of these witnesses to call and tell him the police were inquiring after him. But Brendan was the *only* person Townsend wanted to talk about.

The two men had indeed been colleagues at Monash University. When Townsend joined the gardening staff, he'd noticed morale was low. He was warned by other workers to watch out for Brendan. Some of the gardeners seemed scared of him. A young apprentice soon resigned because she couldn't handle working alongside the man. Brendan threatened his colleagues and if they complained, they reckoned, their car keys would go missing, or they'd find the aerials snapped off their

cars. Brendan was cunning, they said. He knew where the surveillance cameras were and was never caught.

'Got to be careful of Brendan there,' a gardener warned
Townsend in the brew shack, the lunch room. 'He sends nasty
texts, don't you, Brendan? Don't want to upset him.'

Brendan, sitting there, gave a little *tee-hee-hee* laugh, as if
proud of his reputation.

Townsend told the investigator that he'd looked over at him
and said casually, 'Well, Brendan, if you do that – or anyone
else does that to me – your head goes straight through that
door there, see, and that will be the end of that.'

And that *was* the end of that. Everything was fine. The
matter was laughed off, including by Brendan. Townsend had
never personally had any problem with him. Except that he
was always paired with the man, because it seemed he could
actually handle him.

When, a few years ago, Brendan finally left the university, he
considered Townsend a friend and would visit unannounced. It
made Townsend's wife uncomfortable the way Sokaluk skulked
around never looking her in the eye, but Townsend hoped he was
helping the poor sod out, making a bit of a difference. He had to
be in the right mood, though. Brendan talked a 'heap of rot half
the time', tall stories – part fantasy, part brag – spinning off in
never-ending, meaningless asides. He was like a teenager trying
to impress. Townsend would three-quarters listen, answering,
'Oh yes. Really?'

Once, Brendan came around looking for work. Townsend had
an old car on his block, so he said Brendan could cut it up and

take it away for scrap, which he did, leaving not one piece on the ground, not a wire or a single bolt. It was all gone. Townsend had to admit that he did a pretty good job of cleaning up the bush around there.

The bush before it was turned to ash . . .

Ten days before the fire, Townsend had seen Sokaluk lumbering his way and tried to duck out of sight behind some sweetcorn he was growing. Brendan found him and they had a chat.

'Are you still getting married?' Townsend asked. He'd previously heard mention of a romance, but that had apparently ended some time ago. Townsend gave his visitor a bunch of onions he'd just picked and told him he had work to do. Thankfully, Brendan left.

The day after the fire, Townsend found a message on his answering machine, which he now played to the investigator: 'Peter, it's Brendan. I tried to get up, see if you're alright. My car broke down in Glendonald Road and it's torched now. I helped one of your farmer mates last night. Tried to get hold of you, but you were busy. I'll catch up with you later, mate. Hope you're safe and well.'

Townsend hadn't been expecting Brendan, and when he saw his car on Black Saturday he recalled the rumours he'd heard about him, rumours of smoke appearing in places Brendan had just been. His car was parked oddly. Then Townsend saw Sokaluk himself. It was some time before 2 pm. He was getting into the car of a woman called Natalie Turner.

The investigator now visited Turner, an artist, who had been lunching with her parents on Glendonald Road when the fire began. She was introducing them to a new boyfriend, Dane Carozzi. Just after their meal, she heard a helicopter overhead, and looked outside at a mushroom cloud of smoke. She and her boyfriend rushed her children into the car and started to leave, but a blue vehicle that had apparently broken down partially blocked the road. A man stood nearby with his dog, looking dazed. Carozzi urged him to join them: the fire was now perilously close.

'I hope my car doesn't burn,' the man had said when they were finally driving away. He repeated this to himself. When they dropped him home, he said it again, before adding, almost as an afterthought, 'Oh, and I hope nobody gets hurt.'

Turner thought the man said he'd just visited Peter Townsend, although Carozzi thought he told them he was going to see Peter. At any rate, his car had broken down past the Townsend's house, facing away from it. The couple dropped him home.

Natalie Turner calculated she'd left her parents' place around 1.45 pm, placing Sokaluk a little closer again to the time of the fire's start. Turner's trigger for leaving was the sound of helicopters overhead. Paul Bertoncello asked a colleague to look through the CFA data to find out what time fire and police helicopters had been operating nearby.

This was all coming together very quickly, almost too quickly, Bertoncello thought. It had seemed likely the investigators would be in for a long haul, with hundreds of names thrown up, and months spent scouring each blind alley, analysing every red herring. That was how it normally worked. Surely the very first person they were narrowing in on couldn't be the one?

Bertoncello tried to dampen down an instinct saying, *It's him.* It can't be this easy, he told himself, something's not right. He was looking now for the catch, the evidence to eliminate Sokaluk. Each attempt to find it seemed to lock the suspect in further. *It's him.*

Bertoncello looked over his colleague's shoulder as he scrolled through emergency services records. He felt a shiver of recognition. 'Hang on,' he said. 'What's that?' It was the call-line identification data, which listed callers to 000 reporting the fire. Each telephone number included the address at which the account was registered. The address paired with the second caller was 11 Sheoke Grove. It appeared Sokaluk had rung emergency services at 1.32 pm, informing them of the blaze. He must have been right there. Right at the start.

The detectives found a driver's licence photograph of a man with wavy brown hair wearing Coke-bottle glasses. Gawky, a cold sore or scab on his lip.

Brendan Sokaluk had grown up only a few kilometres from where the fire began. He was thirty-nine years old. He was single. He was unemployed and on a disability pension.

For eighteen years he had worked as a groundsman at the local Monash University campus, but two years earlier he'd taken stress leave and hadn't returned. Now he supplemented his pension by delivering the local newspaper, for five cents per copy, and collecting scrap metal. He could often be seen driving the back roads of the Latrobe Valley, scavenging the tips and unofficial dumping grounds for trash he'd haul back to Sheoke Grove.

Later, the police heard about Brendan's odd behaviour from his neighbours. He lived in an estate of modest houses, built in bulk in the 1970s and '80s for power industry workers. One woman, when she moved next door to him, had been warned to keep her distance because he was 'different'. He would stand staring in at her garden, then duck down and hide, not wanting to be seen himself. A few times she found him looking in with a camera. She was with her young child and told him to get lost. Another, older neighbour would be in her living room and glance up, sometimes when she had company, to see him at his fence, silently watching. She already had venetian blinds, but she hung curtains as well.

The woman with the child would hear Brendan banging away in his shed, pulling apart some bit of junk he'd collected. He'd be listening to narrated episodes of *Thomas the Tank Engine* or *Bob the Builder*. The former with their stern morality tales, the latter about high-spirited teamwork. For a while she assumed she could also overhear him talking to a child – his speech was loud, with the detailed rolling commentary she might give her son. It was his dog, she realised. He'd bought the animal when he had a girlfriend living with him, a sweet-faced, guileless woman who seemed to be gone now.

Brendan, it turned out, also liked to go online and chat to people. On his Myspace page, he claimed: 'I'm a young happy male who wants to meet a young loven female to married . . . I don,t read books because they put me to sleep. My heroe is mother earth without her we would all be dead.' On another social media site, myYearbook, he'd posted that he was 'looken for a young wife to shear my wealf with her'. But on the Wednesday four days before the fire, he had logged on and, in the third person, described his mood as 'dirty', because 'no one love him'.

In the past few months, the neighbours had noticed he was lighting more fires in his yard and the fires were getting bigger. They smelt toxic – he was burning the plastic off electrical cords to salvage the copper wiring to sell. People close by had to shut their doors and windows against the choking smoke, which could be so thick, one neighbour claimed, she could barely see her own hand held out in front of her.

The largest bonfire had been lit five weeks earlier, on New Year's Eve. A man at a nearby party saw dangerous-looking flames and went to check them out. Pulling himself up, he spied over the fence a heap of wood and rubbish piled five or six feet high, the blaze a foot or so higher again. Sokaluk was standing right beside it.

The man asked him what he was doing.

'Burning some stuff,' came the reply.

The man from the party called out that it was ridiculous to have a fire of that size in such a small space, but Sokaluk didn't look at him or acknowledge him again in any way. He just stood in the glow of the flames, unmoving.

On the afternoon of Black Saturday, after Natalie Turner had dropped him home, Brendan climbed onto his roof in Sheoke Grove and sat watching the inferno in the hills. His neighbours saw him and noticed that his face was streaked with dirt. He was wearing a camouflage-print outfit and a beanie. One hand shaded his eyes. All around, the sky was dark with smoke. Ash was falling. Tiny cinders burnt the throat on inhaling. Brendan glared down at the neighbours, then went back to watching his mother earth burn.

That night, in the incident room, the police worked until 2 am, finalising the warrants and preparing for an arrest. Dozens of investigators were arriving from Melbourne first thing in the morning and would need to be briefed before targeting different witnesses. Along one wall, the tasks on the whiteboards would soon number in the seven-hundreds, with most finished jobs generating yet more again. Large maps of the Latrobe Valley covered another wall, notated with information on the sequence of the fire's spread. The maps showed all the contours of the Strzelecki Ranges, the plantation forests and patches of national park still marked green, as if from a bygone era.

To the detectives, nervy with adrenaline, everything now seemed like a sign. On Tuesday night another fire had been lit in a park not far from where Sokaluk lived. Sitting in the incident room, they now half wondered if it was their suspect attempting to throw them off the trail. Did he want to point them in the wrong direction, or were local kids just up to something stupid? After all, what was there to do here on a hot February night with the whole town on edge?

According to the listings in the paper Sokaluk delivered, there was always a group meeting somewhere in the region.

Even if you had to drive for miles, there were clubs for music lovers, orchid growers, Scottish dancers, chess players, bereaved parents, amateur astronomers, vintage car buffs, the Coal Valley Male Chorus, stamp collectors, knitters and quilters, those with asbestos-related illnesses, or Down syndrome, or arthritis, or Alzheimer's. There wasn't much for young people to do, though. The police patrolling the streets found them empty. At a glance, the only place open was the power station, whose turbine hall in the distance was like a giant light box, the chimney towers above palely gleaming.

Nights here were always like this. A long dark wait until morning, not knowing, sometimes, what you were waiting for. The quality of the silence outside had changed over the years. It hadn't always had this sharpness to it. Churchill had been grandly conceived as a utopian hamlet, with parks, a cultural centre, a theatre, a department store and hotels. But things hadn't quite worked out that way. None of those buildings were ever erected, and instead the centre of town was the supermarket car park. After the privatisation of the state's energy grid, a high percentage of the 4000-strong population were unemployed. The only reminder of the original vision was the cheap-looking cigar rising out of the ground.

Brendan's house wasn't far from this monument, and to prepare for his arrest, plainclothes officers had driven and walked past his place to scope it out. Analysts had undertaken habitation checks to confirm that he paid the rates and utility bills.

Here, each street was named for a different tree. Sheoke Grove ran between Grevillea Street and Acacia Way, which connected to Banksia Crescent, Coolabah Drive and Willow Street. The hawthorn, elm, wattle, cedar, hakea, blackwood and manuka were also on the map. They were addresses to conjure, for the power workers, all that blooms and is natural. It was a place for a fresh start. But now the street names listed trees that had recently been torched, and it was hard to resist the symbolism of that mad golden cigar. The flame and the timber just waiting for each other, built into the town's very design.

Brendan Sokaluk's dark-brown brick bungalow had been made in the same severe mould as others in the street, but while his neighbours had clipped their hedges into careful spheres and hung nylon lace in the windows, Brendan had not tried to soften the stark edges. Summer had killed off everything but an orange tree in a dusty bed. It was the only sign of life as Detective Senior Sergeant Adam Shoesmith knocked on the front door. The drab curtains were drawn. Not a sound came from inside. Henry and Bertoncello stood beside him in the late afternoon sun. The policeman knocked again. Silence.

Bertoncello walked down the side. In the backyard, amidst

the empty garden beds and the rusting Hills hoist, lay the junk. Every square foot was stacked with it – wire, piping, flyscreens, components of old office chairs. The detective was used to visiting run-down addresses – disorder was the recurring theme of most houses he called on – but this was different. This was a private junkyard.

Bertoncello tried the back door. Again there was no answer, and it flashed through his mind that they may have arrived too late. If Sokaluk had been tipped off about a police visit – or even if he hadn't, but had lit the fire and been overcome by the horror – he might have taken his own life.

Now the other two men joined Bertoncello in the backyard. None of them knew much about their suspect. Fearing for his safety, they were moving faster than they'd have chosen, and they stood taking in the scene. An incinerator constructed from a 44-gallon drum sat in the centre behind a rickety wall of odd bricks and timber. Here, and in the front yard, patches of the ground were burnt. Perhaps Brendan Sokaluk couldn't help himself.

Bertoncello found more junk inside a shed-cum-workshop. A pyramid of it in the middle, like a shrine to the seemingly useless: old machinery parts, cords, rusting bumper bars, bits of televisions and whitegoods, bike and car wheels, an aging air-conditioning unit, car doors stripped of their metal handles. This hoard was layered on top of wheelie bins crammed with other abandoned objects. All that was missing was their collector.

On the fly, the detectives made a new plan. They had an officer back at the Morwell station call Sokaluk's mobile phone and pose as a consultant from his car insurer, claiming she needed him to sign some documents and asking if he was home. No, he was on his paper round, not far away in Maple Crescent.

In the unmarked police car, the detectives wound back through the circular streets, a perfect blue sky and transmission towers overhead. The power workers who'd bought here had aged, but despite the rusting letter boxes and peeling eaves, all obeyed a standardised neatness. The eye-shaped petals on the curtains kept the neighbours in check.

In Maple Crescent, a pudgy man was pushing a removalist's trolley stacked high with newspapers. A terrier was tethered so close to the side it was half walking, half being dragged along. The dog's owner had a peculiar gait, an awkward kind of bounce to his step. Wearing a sweatshirt, shorts and long socks, all a shade of faded green, and black velcro sneakers, Sokaluk looked like an overgrown kid in his school uniform. A little bored and in a hurry.

Adam Henry swerved onto the nature strip, cutting him off. When the detectives got out and were closer they could see their paperboy had aged since his driver's licence photograph had been taken. He wore his greying hair in a buzz cut, shorn close to the skull. There were fine lines under his eyes.

Henry showed his police identification. He asked the suspect his name and date of birth – 'Brendan Sokaluk', '11 of ten, 69' – and told him he was under arrest for arson.

Brendan didn't tremble like most people in that situation did. Bertoncello took hold of his arms, handcuffing him, and there was no quaver. Nor did his face reveal fear. As he was read his rights, he was almost blasé. The headline repeating on the pile of newspapers read *RAZED*. Pictures of the fire's damage wrapped around the front and back pages. Whichever way Brendan had folded the paper into letterboxes, there would have been views of a black and white world, police crime tape blowing in the breeze.

His denial was matter-of-fact: 'I didn't light any fires,' he said, and his speech was slurred, rather in the way of a deaf person. 'But I'll put my hand up, like at Monash.'

The detectives didn't stop to ask what he meant. They bundled him into the car and left the trolley of newspapers on the street, undelivered.

Back at Sheoke Grove, Sokaluk was served with a search warrant while still in the car. Then Henry and Bertoncello waited there with him, making small talk. Usually this was a soothing kind of patter to calm the accused person down, but Brendan wasn't overly interested in being soothed. The detectives sat there conscious that Shoesmith was now entering this man's front door, coordinating with the other officers who'd arrived to video the property before the forensic team swept through.

It was dark inside the house with the curtains drawn, although light came through the gaps where they had half fallen down. When Shoesmith's eyes had adjusted, he saw that the layout was compact. The front door opened onto a small vestibule with an adjacent living room. Someone, at some point, had tried to brighten the place up by painting the walls mint green, and the architraves and woodwork a gaudy, high-gloss pink.

A narrow hallway led to three bedrooms. One of these was furnished with old gym equipment: a running machine, a stair-climbing machine, a bench press with some weights. But workouts had apparently been abandoned: the equipment was covered in bags of clutter, including an old pack of Celebrity Slim, a seven-day meal replacement plan.

In Brendan's green bedroom, an iron bar rested against the wall near the bed, as though for protection against intruders. Three old double mattresses were piled atop the metal frame of a single bed. The sheets were faded, stained. An oversize teddy bear sat in the wardrobe amongst the camouflage-print clothes.

Another room, with what looked like salvaged office furniture, had blush-pink walls, a wardrobe the shade of a ballet slipper. On the desk sat a computer surrounded by losing TattsLotto tickets, the registration of the now burnt car, a business card from Connolly's towing firm, a computer magazine, and a drink coaster with a bare-breasted blonde posing on a beach, captioned *The Heat is On!*

When Sokaluk was taken out of the police car he was led up the driveway to the back door, as he'd requested, so his neighbours wouldn't see. It was cramped inside the house with the small crew of officers. Bertoncello removed Brendan's handcuffs and asked for his mobile phone.

Shoesmith had set up the camera tripod for a video interview, just by the front door's pink frame. Above them, light bulbs without shades were stuck into the ceiling sockets.

The interview was a formality before the police executed their search warrant. Shoesmith had conducted these conversations countless times, but not usually for such high stakes. He was trying to keep the mood calm – his own and the suspect's – collegiate even, as if they were just standing around having a friendly chat.

All he wanted to do was to keep this arrest as quiet as possible and get Sokaluk safely out of town. Arson had never been so publicly discussed as a factor in major bushfires, and Black Saturday's were the worst in living memory. The people of the Valley, full of grief and rage, agreed with the prime minister, Kevin Rudd, who'd appeared on television a few nights earlier. 'What do you say about anyone like that?' he asked, shaking his head in disgust. 'There's no words to describe it . . . other than it's mass murder.'

The detective pressed RECORD.

Shoesmith: 'Alright, it's Detective Senior Sergeant Adam Shoemith of the Arson Squad . . . Ah, we're currently at the address of number 11 Sheoke Grove in Churchill. It's Thursday the 12th of February, 2009. The time is now, Brendan, do you agree the time is now twenty-two minutes past 5 pm?'

Sokaluk: 'I suppose so.'

Shoesmith shows Sokaluk his watch.

Sokaluk: 'Close.'

Shoesmith: 'Um, to my right is the occupier and owner of these premises, Brendan Sokaluk. Brendan, can you state your name, age and date of birth?'

Sokaluk: 'Got (inaudible). Brendan. Age, um, thirty some-thing or other. What's the other bit?'

The detective is thrown. This man's reactions seem slower, more disjointed than they were when the camera was turned off. Brendan asks him suddenly who the 'bad man' is, pointing to one of the police officers who is waiting outside.

Shoesmith: There's no bad man . . . Can you remember what you were doing at the time when we first approached you?'

Sokaluk: 'How many takes are you going to take of this?'

Shoesmith: 'Sorry?'

Sokaluk: 'How many takes?'

The detective tries to get things back on track by reminding Sokaluk he was doing his paper round before they escorted him home.

Shoesmith: 'And you had your small dog with you?'

Sokaluk: 'My doggie.'

Shoesmith: 'Yes, is that right?'

Sokaluk: 'Yeah, my dog.'

Shoesmith: 'The policemen informed you that you're not obliged to say or do anything, but anything you say or do may be given in evidence. Do you remember that?'

Sokaluk: 'Not really, didn't take much notice.'

Shoesmith: 'Okay, do you remember that they told you what the offence was that you were under arrest for?'

Sokaluk: 'I'll say yes, but I don't remember.'

Shoesmith: 'Do you agree that that was in relation to the Churchill fires on Saturday, and the offence of arson causing death?'

Sokaluk: 'Probably.'

Shoesmith: 'Does this all sound familiar or you're not sure?'

Sokaluk: 'Ah, I don't really comprehend things too much.'

Shoesmith: 'Well, do you agree that these things that I've stated to you happened? I mean, it's only happened in the last sort of half an hour.'

Sokaluk: 'I'll say yes, just to make you happy.'

They suspended the interview. It had lasted barely six minutes. There was stunned silence. Only moments earlier the detectives believed they'd been holding a relatively normal conversation with Brendan. On camera, he appeared to have difficulty understanding basic questions and his speech was suddenly harder to decipher. It was standard to be lied to as a detective, and Shoesmith had witnessed various ruses by suspects to gain advantage, but never this: this man seemed to be taking on the role of village idiot, twitching and incomprehensible. The detective felt he was watching a bad actor. If the crime hadn't been so serious, he thought later, he'd have burst out laughing.

At the moment of Sokaluk's arrest, teams of detectives had been dispatched to take witness statements. Shoesmith wanted to give people as little time as possible to confer and potentially adopt each other's stories. The detective interviewing Sokaluk's father now sent back word that the suspect apparently had an acquired brain injury. There was no information about how it had been acquired, and at this stage Shoesmith was inclined to think that, behind the acting, the accused was not so different to a lot of the guys from this part of Gippsland the police chased around. He decided to get Brendan back to the station and wait for a specialist to come and assess him.

A small woman with bobbed dark hair had swept into the driveway and was giving Inspector Ashworth a piece of her mind. It was Brendan's mother, who had arrived to check that her son was alright, and now watched him being led in hand-cuffs to the car.

'Don't bash Mum!' he called to the police officers waiting to search the premises, as they stared back straight-faced.

During the ten-minute car ride, Sokaluk seemed to again act relatively normally. Perhaps he had some cognitive issue, but they could at least understand what he was saying, and it veered towards the cocky.

'Two hands on the wheel for beginners,' he said to the officer carefully driving. 'Are you nervous, mate?'

At the station, he was placed in an interview room with Paul Bertoncello, who acted as if the two of them just happened to be there and it could all be a misunderstanding. This was the warm-up for the actual interview and for the next hour, the detective tried to establish some rapport with the man under arrest.

It's an art, persuading people to speak to you. In subtle and unsubtle ways, detectives try to sway their suspect towards them. Every gesture, every variation in tone, is part of a calculated exchange, a set piece, often with well-worn nuances and moves replicated constantly on police procedurals: the handcuffs clicking on and off to show dominance, providing a special drink or a smoke to build complicity, offering sympathy to imply that in any other situation everyone would be great mates. The detective becomes a kind of confidence man, with the same worldliness and steely patience.

Moment by moment, Bertoncello attempted to advance the relationship, and all the standard techniques fell flat. This man, with no criminal record, inexperienced in the ways of the police, should have been more susceptible to opening up. But Sokaluk had an insouciance that in the circumstances was galling. Perhaps he shared the local mistrust of outsiders.

'Do you like sport?' the detective asked. No, he didn't, at all. 'Do you have a girlfriend?' No, and Bertoncello got the impression he didn't have a high opinion of his last one. Sokaluk mostly just wanted to know where his dog, Brocky, was, while Bertoncello kept assuring him that the animal was safe.

Back at Sheoke Grove, the detective had scanned the room for personal details. There were a few books on a small set of shelves: *Australian Gardening A–Z*, *Your Gardening Questions Answered*, *Herbs* and *Home Landscaping*. (Ironic, given the state of the yards.) But Bertoncello had also noticed a complete collection of *Star Trek* DVDs, and so he now spent as long as he could manage chatting about Starfleet and the Enterprise.

Time didn't pass easily.

In the incident room, Shoesmith and Henry were being updated on what the investigative teams had discovered. As new witness statements came in, the detectives collated them to establish a timeline for Sokaluk's movements on the day of the fire. A forensic team had seized his car from Connolly's and were testing it for evidence. Another team swept through his house, taking out the camouflage clothes and photographing each item. They found matches in the bathroom and a lighter in the bottom drawer of the wardrobe in Brendan's bedroom: these were photographed too. So was his incinerator and a butane torch in the shed full of junk. His computer was carried out to a police vehicle, and an electronic crimes analyst waited at the station to pull it apart.

Before the interview started, the police wanted the best possible picture of who Sokaluk was and what he'd done before and after the blaze. Ideally, more information would have been gathered prior to the arrest, but Shoesmith was too worried about vigilantes to leave Sokaluk walking the streets with his dog and trolley. And so, in truth, the Arson Squad had next to nothing. Their case was completely circumstantial. Only Sokaluk's dialling 000 linked him to the start of the fire. And that, like everything else, could easily be explained away by a defence lawyer. Shoesmith could almost hear a barrister dismiss as coincidental Brendan's driving around the bush at the time of ignition. His sitting on the roof and watching the inferno? Well, they were just the odd habits of a man who was a little unusual.

When Adam Henry felt he had a viable interview plan, he joined Bertoncello and their suspect in the cramped room. He hoped this man had grown comfortable with Bertoncello and would be prepared to talk. Three tapes were inserted into a triple-deck tape recorder, a master and two copies. It was now 6.48 pm.

Henry asked Sokaluk for his full name.

He apparently couldn't remember his middle one. 'Starts with J,' he offered blankly.

The officers were wary. It immediately seemed, once again, that Sokaluk was speaking less clearly and acting more disabled than in the previous few hours when he wasn't being recorded.

Henry tried repeatedly to read the accused his rights, and then to check if he understood them.

Henry: 'You don't have to say or do anything, but anything you say or do may be given in evidence . . . What does that mean?'

Sokaluk: 'Keep your mouth shut, does it?'

Henry: 'I don't know. You tell me what it means?'

Sokaluk: 'I don't know what it means —'

Henry: 'Alright.'

Sokaluk: 'Take a guess.'

Henry: 'I'll break it down into pieces. You don't have to say or do anything. What does that mean?'

Sokaluk: 'Be quiet, does it?'

Henry: 'But do you have to talk to me?'

Sokaluk: 'You are a stranger.'

Henry: 'Do you understand what I mean? Do you have to speak to me during this interview?'

Sokaluk: 'So, I have to, wouldn't I?'

The detective tried again to explain Brendan's right to silence.

Sokaluk: 'So, I just sit here and be quiet?'

Henry: 'You can if you want to.'

Sokaluk: 'Right.'

Henry: 'Okay, so what does that mean?'

Sokaluk: 'That youse told me to sit down and be quiet.'

Henry: 'No, I'm not telling you. It's up to you what you want to do. Do you have to speak to me during this interview?'

Sokaluk: 'You said I didn't have to.'

Henry: 'No, you don't have to if you don't want to. But if you want to, you can. Do you understand that?'

Sokaluk: 'Yeah, I like to talk . . .'

Henry: 'Okay, alright, now anything that you do say to me, okay? Say or do, can be used in evidence. Do you understand that?'

Sokaluk: 'What they, like, twist around and use against me?'

The exchange felt ever more ludicrous. Henry believed Sokaluk was deliberately being a smartarse, acting vague, playing up whatever his impairment was to disrupt the interview. After seven minutes, the detectives stopped the tape. Were they imagining it, or did the accused again recover his more capable demeanour once the recording was off?

It was perfectly legal for the Arson Squad to interview Brendan prior to him being medically assessed, but they just weren't getting anywhere with the soft, friendly approach. Paul Bertoncello decided to tell his superiors, Shoesmith and Ashworth, they had the wrong interview team: the questioning should start again with officers using a less sympathetic register.

It would be more than two hours before a forensic medical officer arrived from Melbourne to determine whether Sokaluk was fit for the interview process. The senior officers had Brendan transferred to the watch house, where he could be remotely monitored. Being locked up might unsettle him enough to jar him out of the game he seemed to be playing. Bertoncello and Henry were directed to have a break and then continue, but to be firmer.

At 8.30, the two detectives headed out into the dusk looking for some dinner. They'd barely eaten all day. Down the hill, the power station's lights were glittering for the night shift. The pair walked in the other direction, a few metres up to Commercial Road, Morwell's main drag. On one side, there was a train track; on the other, the number of vacant shops with FOR LEASE signs lent an air of desolation to such life as the street had. The only place open was a Subway, where the men ordered food.

Bertoncello and Henry had known each other only vaguely until now, but for the next three and a half months they would stay in the same motel, work the same long hours, eat the same takeaway. The clock didn't really matter in these jobs. Living away from their families, with no real reason to leave the make-shift office, they ended up putting in sixteen-hour days. Everyone shoulder to shoulder in the incident room, forty investigators at the peak, dealing with the enforced intimacy and the inevitable clash of moods and personalities.

The past twenty-four hours had been exhilarating, but the frustration with the interview opened up others. Henry's wife was at home in Melbourne with their newborn, who had arrived early and was only a week out of intensive care. Bertoncello's wife was caring for their four kids, all aged under ten (and all of whom they'd struggled to name, because each possibility reminded him of some delinquent he'd arrested). Being a detective did not always make for an easy family life: the divorce rate amongst police is double that of the rest of the population.

While they waited for their food, a message came through: Sokaluk urgently wanted to speak to them.

The detectives rushed back, scoffing down their enormous sandwiches in the twilight. They believed they'd twice witnessed the suspect behave differently on tape, so this time Adam Henry would wear a micro-recorder into the cell.

'What's up, mate?' he soon asked.

'I want to start talking,' Sokaluk replied.

To ensure that later it didn't appear Brendan had been

tricked into speaking, Henry decided to once more try formally interviewing him. He was brought from the watch house back to the interview room. It was nearly 9 pm.

In the video recording of this interview, Brendan is sitting in a corner by a laminex table. There's no natural light; a grey stripe winds along the claustrophobic walls. His shoulders are rounded, his belly protrudes from a baggy, stained sweatshirt.

On the other side of the table are the police.

When Brendan is read his rights and asked to explain them, he seems to better understand this time, although only slightly: 'If I don't [speak to you] I have to go back [to the cells].'

Henry asks him repeatedly if he wants a lawyer.

'I don't know any.'

'We could make some arrangements, if you want to speak to a lawyer, that's up to you.'

'There wouldn't be any lawyers around at this time of night,' he slowly answers.

'We could try.'

'No, because it would be wasting people's time.'

Eventually Henry says, 'Okay now, you said you wanted to talk to us?'

'Yeah.' Sokaluk speaks every word in the same dull, flat tone; a *shh* sound for *s* and *f* for *th*. 'Want to tell you what happened Saturday, regards to the fire stuff. First, I was smoking in the car when I was driving.' His sentences run together; only some are completed, and these fragments sound slurred. He explains that to, 'get to my mate's place, you can go the bitumen road or the gravel, and the bitumen road is dangerous because of hoons.

So, I go the gravel road, and I like to take shortcuts off the gravel, but when it's rough it just shakes the car. And I was smoking, a bit fell down and so I grab a bit of paper to grab it and flick it out sort of thing, have to squish, flick that and it must've ignited. And I went up this track [Jellef's Outlet], this road there that goes up; top of this was rubbish up there. I went up there and I reckon the car wasn't working too well and stuff, and had to turn around. And then I noticed there was fire and I panicked, and I called ooo and I just tried to get away as quick as possible, just panicked.'

Here it is: the Arson Squad now have a confession, of sorts.

The undercurrent in the room changes. The detectives hang on every charged word. Large swathes of this strange story are delivered without any eye contact. Very little animates Brendan's face, but he runs his fingers repeatedly along the top of the table and below it, again, again.

Sokaluk: 'I haven't been able to sleep properly after this. I like the bush, just didn't want it to go up . . . Makes me sad inside.'

Henry: 'Why is that?'

Sokaluk: 'Because people died, was my fault and I have to put up with that.'

Henry: 'How was it your fault?'

Sokaluk: 'Because I was stupid. I burnt down one thing I loved in my whole life is the forest and my stupid actions stuffed it up. Now I have no place up the forest, going to sit and watch the fish, look at the creeks. Because my life gets hard from the stuff, and stress wise and that, I would go up there and sit, and this would to relax, I found. But now I've destroyed all those areas and all those poor people died, so stupid. That's why I always got into trouble at Monash, because I was stupid all the time, do things, after forget. I didn't want my, my friends to get hurt. Next day I try to forget about it, try to . . . going to wish it would all go away because I want to be able to have some proper sleep.'

There is no provision for the senior officers to remotely monitor this interview. They can't hear the pidgin-like sentences Bertoncello and Henry are deciphering. If Shoesmith wants to send them word, he has to slide a note under the door and wait for them to discreetly retrieve it. The detectives have previously decided to let Sokaluk give his account of events and to keep their questioning to a minimum. The more he speaks, the more he might incriminate himself.

It is clearer to them that Brendan is not quite right. His speech is childish, with words missing, syllables mispronounced, and it is all delivered in the same expressionless manner. They also figure, though, that he owns his own house, he's apparently paid the bills, and can use a computer. He'd been canny enough to realise, in the cells, that his name hadn't been pulled out of a hat. He then presumably calculated how much they knew about him – which isn't much – and, in giving his story he is now the one leading the interview. When, very occasionally, he darts a look in the detectives' direction, it's to appraise them. His eyes move as if behind a mask.

At 9.14 pm, they stop talking. A forensic medical officer has arrived from Melbourne to assess Brendan. She takes his medical history and determines that he is fit to be questioned on the condition an independent third person is present.

The interview recommences at 11.09 pm, and in the video recording, a craggy-looking justice of the peace sits next to Brendan. This man has no specific training for the role, although he has spent time with some intellectually disabled people at the church he attends.

Henry, conscious that the accused may have such a disability, reminds him again of his rights so that later there will be no legal contest.

Brendan's hands are clasped in front of him.

Henry: 'What's your full name?'

Sokaluk: 'Brendan J. Sokaluk.'

Henry: 'What does the J stand for?'

Sokaluk: 'Can't remember.'

Henry: 'Okay. Is it fair to say it would be James?'

Sokaluk: 'Could be.'

When Henry says, 'I intend to interview you in relation to the offences of arson causing death and intentionally causing a bushfire,' Brendan doesn't react at all. He doesn't meet their eyes. After the caution is explained again, he yawns.

Henry repeats that he has a right to a lawyer and Brendan looks up briefly, then looks away, strenuously yawning. He says for the second time that he doesn't know any lawyers.

Henry asks if he knows what lawyers do.

Sokaluk: 'Goes to court . . . defends people.'

Henry: 'What type of people?'

Sokaluk: 'Bad people.'

Henry: 'Not necessarily bad people, people that have been charged.' He offers to get a phone book and help find one.

Sokaluk: 'No, it's too hard.'

The independent third person looks as blank as the accused. Sometimes, in the station, there were jokes about these individuals needing an independent fourth person. This one doesn't suggest they open the book at 'L' and ring around.

'This is your choice,' Henry says.

'Keep going,' Sokaluk replies.

The detectives ask Sokaluk to take them through his movements last Saturday. He tells them that it had started like all his Saturdays did: 'went out and shopped with my Dad and stuff'.

His father, Kaz, short for Kazimer, had told the investigators who interviewed him that this was their regular routine. They drove to Morwell, where they visited the TAB to lay some bets, then drove to Traralgon, where, both being passionate about cars, they went to Autobarn and Supercheap Auto. (Police would later locate the surveillance videos, which showed two stocky men, father and son, doing their morning rounds. Kaz, bearded, with a tattoo stretching up his arm, was an ex-mine worker, having been pensioned off with a bad back twenty-five years earlier.) On their return, they stopped at KFC for an early lunch, then at the Mid Valley Shopping Centre, a mall outside Morwell, where Kaz paid twenty dollars off his lay-by of DVDs.

All this time, Kaz had told the investigators, 'Brendan's car was running a bit rough, the fuel was evaporating and it was trying to stall. In the heat, it was sounding like a tractor . . . really chugging.' As they returned to Churchill, they saw a small crowd in a paddock watching a grassfire. 'People are like that, aren't they?' Kaz said. 'Got to rubberneck.' They stopped to rubberneck too: the Morwell fire brigade was on its way, and this was a form of public theatre.

Afterwards, as Brendan dropped his father home, he mentioned he planned to drive into the hills. Kaz knew Peter Townsend lived that way, and also an ex-neighbour, Dave. It was now forty-four degrees and the car had no airconditioning. Kaz told him not to travel further in the heat, and offered to tune the car up when the temperature dropped.

Instead – it slowly emerges, these facts coming from the accused in a blur of random detail – Brendan drove to his house a few streets away and changed into sturdier boots. He then headed around the corner to the petrol station on Acacia Way, where he bought a packet of Pall Mall Slims Green – 'Pally Wally', as he called them.

While he was driving, he now tells the detectives, he must have taken a cigarette out of the packet with his mouth, and then lit it with a cigarette lighter. The one in the car didn't work.

He drove towards 'my mate's place', along Glendonald Road, travelling at a 'dawdle' because 'we' – he and his dog, Brocky – 'wanted to spot the wildlife', and search for rubbish. He'd stuck to the dirt service track rather than the gravel road, because further up it was bitumen, and 'people roar down there real fast and they get airborne'.

The track's uneven surface made the car unsteady, he claims. 'The vibration rocks the car about, and stuff, and so I was smoking and I had a burnt bit [of the cigarette] fell off onto the floor.' He used a serviette, 'fast food paper', to pick the ember up. 'I squished it out sort of thing and when I threw the paper on

the road it ignited. I didn't know. It was too late. I panicked . . .
I called triple o and started telling them there was a fire on that
road. I did a bad thing and I'm scared shit, shit-scared.'

For arsonists who thrill to the drama of the emergency response,
it is not unusual to call in the fire they've lit. Later, police specu-
lated that Sokaluk may have hidden on a track in the overgrown
plantation and waited to watch the firefighters arrive at the scene.

Geoffrey Wright, who had been trying to catch his wife's
horse further up Glendonald Road, was confident that while the
fire engine was pulled up, warning him and others to evacuate,
the sky-blue car drove past on the engine's other side, before
breaking down further along the road. The CFA crew all believed
the car had already stopped when they arrived. They remembered
seeing Brendan in the disorder, standing in his camouflage-print
clothes, staring at the blaze as it approached.

Through the truck's window, one volunteer called, 'Get out of
here, there's a fire!' Then, when Brendan didn't move, 'Are you
here to help someone?'

He didn't answer: he just stood cradling his dog, watching the
flames.

The police knew that when Natalie Turner and her boyfriend
had pulled up he got into the car with them. Brendan had been on
the phone to his father during this ride into Churchill. The couple
heard him telling Kaz about his car's breakdown, although not
that he'd just reported the fire.

A few hours later – after his neighbours had observed him
watching the blaze from his roof – Brendan decided to walk back
towards the fire, perhaps to check on his car. His father and he

had completely rebuilt the vehicle together, and Kaz had told him not to drive further.

'My old man was angry,' Brendan says, 'because I torched the car, he was very upset. I think he hates me.'

Brendan walked past the university where he'd once worked. Two of his former colleagues happened to be standing outside, listening to what they presumed were houses exploding, when they glanced up and saw him trudging past. He'd assiduously avoided the place since he'd left, and they were surprised to see him.

'I was a landscape gardener at Monash,' Brendan explains, 'and I ended up being their punching bag.'

The accused didn't acknowledge these men as he went by. He hiked back towards the fire through the paddocks, bypassing the police roadblocks. Later, different volunteer firefighters confirmed seeing a man in the dull, smoke-filled light, walking a dog in the midst of the raging fire. As one volunteer noted: 'There was truck engines going, there was radios going, there was the quick fill pump going, there was noise . . . it seemed very out of place that someone would be taking a casual stroll with their dog.'

Brendan tells Henry and Bertoncello he went to check on his friend, Peter Townsend, who wasn't home (a claim that turned out to be untrue, as Townsend had returned to his property to try to save his livestock). Instead, Brendan found himself across the road at the Fergusons' place, helping them save their house, until Tony Ferguson asked the Hazelwood North fire brigade to give their unexpected guest a lift home. On the way he went with the brigade to the site where they were pumping water. The dam was on a property with a Confederate flag strung up outside, along with a threat to shoot trespassers on sight. The owner, who according to Brendan was 'a psychopath, real mean and nasty', 'getting drunk and . . . popping tablets all the time', told him he'd shoot whoever had lit this fire. Brendan kept quiet.

A firefighter working near the man's dam had told police: 'This guy was just walking around, watching us.' He said Sokaluk was holding his terrier so its feet wouldn't touch the burnt ground. Eventually a CFA officer asked a passing car to take him home, where, Brendan says, he 'couldn't sleep with what happened. Not happy, I was sad . . . upset over it or something. What a stupid thing I did and I was shit-scared to tell anybody.'

He keeps yawning now. It's getting close to midnight. There's static in the long gaps between question and answer. He starts picking at a scab on his neck.

'Alright, so you're not an active firefighter, are you?' Henry asks.

'No.'

'No?'

'I don't really like fires,' replies Brendan, deadpan.

Pressing what he takes to be his advantage, Henry says: 'You sound like you've had a bit of experience with fire before?'

The atmosphere in this grim room quickens.

'When I was little and that, I was in the fire brigade and that,' Sokaluk tells him.

'So, I take it you're not in it anymore?'

'No.'

'Why is that?'

'There was a bad cop in Churchill and he had me kicked out.'

'Okay, so you're quite familiar with fire and firefighting?'

'Don't remember it all.'

'Sounds like you're pretty handy?'

'Know how to help people and stuff like that . . . Put it out.' Sokaluk yawns again and it's the only time his face seems to change expression. His knuckles run along the edge of the table. His short nails are lined with dirt, there's grime ingrained on the whorls of his fingertips.

That he has had some association with the CFA interests the detectives. Although statistically it's uncommon for firefighters to deliberately set fires, it *is* common for arsonists to be firefighters. Volunteering to battle local blazes offers camaraderie and status. It's a bonding, adrenaline-filled service, for which politicians and the media turn some of those in the ranks into heroes. And, of course, if there are no fires when the season starts, someone feeling powerless and forgotten might start to itch for the thrill.

'What about fire behaviour?' Henry tries.

'Wouldn't have a clue,' replies Sokaluk.

'How does fire travel?'

'That one went away too fast.'

'What are some of the things that make fire travel?' Henry asks.

'Stuff that can burn.'

'Alright,' Henry says, 'what else helps fire?'

'Probably wind.'

'Is it quite common for fire to travel up hills or down hills?' The blaze had been lit at the pit of a natural basin, the incline all around accelerating its spread.

'It goes where it wants to,' the accused answers pragmatically.

At 12.29 am they have a 25-minute break. During this interval, the detectives are given a copy of a Crime Stoppers report. The digital imprint of his confidential statement had been found on Sokaluk's seized computer. Two days earlier, he had submitted a form headed *What is your information about this crime?* The form had various subheadings, also in the form of questions.

When the interview resumes, Henry reads them and Sokaluk's answers aloud:

What is happening? a bad man lighten fires
When is it happening? on saterday
Where is it happening? glendonal road outside Churchill
Why are they doing it? on edge of plan tastion
How are they doing it? can,nt see his back is to me
Is there anything else you believe may be helpful? its a d,s,e [Department of Sustainability and Environment] fire fighter lighten a fire why is he doing this bad thing 1 could of died if the wind chance 1'd tryed to tell the police but they were to byse
Who is committing the crime? 1 saw a d,s,e fire fighter light a fire on the edge of the road as 1 went pass 1'm sceared that the bad man will get me

There's a silence after Henry finishes.

'They said it was anonymous,' Sokaluk offers. Any contrition appears minimal. His tongue is moving around in his mouth like he's cleaning something stuck there.

Putting in a false Crime Stopper's report to cover for having dropped a cigarette is only a summary offence. The detectives are still a distance from proving arson causing death. Henry tries something else: he explains that analysts have taken Sokaluk's computer from his house and have been examining it. For the first time, the accused appears uneasy; he starts shifting back and forth, rubbing his legs.

Henry: 'What have you been doing on your computer?'

Sokaluk: 'I play games and go on internet.'

Henry: 'Yep, and when you go on the internet, what do you look at?'

Sokaluk: 'Lots of stuff.'

Henry: 'Like what?'

Sokaluk: 'Porn, normal porn sometimes.'

Henry: 'Normal porn?'

Sokaluk: 'Yeah . . . Only look at normal stuff.'

Brendan is given another caution.

Henry: 'Now tell me a little bit more about these porn sites?'

Sokaluk: 'I don't know what it's got to do with the other, the fire.'

Henry: 'Would it be fair to say that there's underage people on those sites?'

Sokaluk: 'Couldn't say. I don't remember. It would have been a long time ago that, was it? A bit hard trying to get some of that rubbish off, people tell you these things and you go on and then look at it, like normal ones, and all this other shit pops up.'

He claims this 'bad stuff' or 'naughty shit', the child pornography, would appear when he was trying to download games, or look at 'normal' porn – after all, he says, 'most people have got pornograph on their computer'.

Henry moves on, asking the accused if he's looked at anything related to Black Saturday on the internet.

'It was too depressing,' Sokaluk says. 'It made me very upset.'

In his search history, the forensic team have found dozens of photographs of firefighters battling the blaze. One was of a fire engine streaking through flames and smoke; the smoke wasn't just towering, it was terraced, a whole horrific kingdom drawn in every shade of brown and black. A dense, volatile pyrocumulus cloud, stretching up several kilometres, colonised the sky.

Bertoncello sits silently next to Henry. He is the junior officer. Less fazes him each year, but he's unnerved now. The term 'cold-blooded' keeps coming to mind. He feels Sokaluk is saying what he thinks he ought to say, what he thinks they want to hear, but there is no emotional resonance to the words. He gives an answer that implies remorse in the same register he's used to ask about dinner.

Bertoncello hadn't known, when he joined the police service – no one does – the scale of the damage he'd go on to see. One minute he was a teenager watching *Police Rescue*, where an actor in a helicopter was winching another actor to safety, thinking, That looks a good job; the next he was learning a daily lesson about people's ability to inflict pain, and the stories they then tell themselves and others.

Who was this man they'd arrested? How smart or dumb or cunning or oblivious he was seemed to change from moment to moment. The detective didn't claim to know about the diagnostic subtleties of pyromania, let alone sociopathy, although there was something in Brendan's lack of empathy that brought this word to Bertoncello's mind.

When he takes over the interview, Bertoncello tries combinations of questions in the hope that they might unlock some explanation of why, as he suspects, the man opposite him deliberately lit an inferno.

Bertoncello: 'How did it feel fighting that fire?'

Sokaluk: 'Horrible. It was hot.'

Bertoncello: 'Was it exciting?'

Sokaluk: 'No. It's frightening. It's scary.'

Bertoncello: 'Have you tried to fight other fires?'

Sokaluk: 'No.'

Bertoncello: 'Have you seen other fires up close?'

Sokaluk: 'No, I don't want to see anymore.'

At 1.25 am the detectives are about to suspend the interview. Henry asks one last question. 'Have you told anyone what you did?'

Sokaluk replies, 'No, 'cause they, people, would go and firm a lynch thing and chop a person up, so I kept quiet.'

Sokaluk spent the night in the cells, and the next morning, in a video recording of a field interview, his khaki uniform of sweatshirt and shorts is crumpled. He's standing on Glendonald Road, not far from where his car had broken down, in his black velcro sneakers. His hands are in his pockets, fidgeting. He's dark around the eyes, and doesn't meet the gaze of the detectives or the videographer.

Shoesmith, Henry and Bertoncello stand in a line next to him. They look as sombre as undertakers in their white business shirts and striped, police-issue ties. The camera pans around the empty road, still blocked by a police barricade. There's a strange silence. In the wind, the leaves of unburnt trees whip the air. Suddenly, an approaching fire engine can be heard. Nearby there's been a flare-up.

'Oh, great, fire truck,' Sokaluk mutters, lifting a hand to shield his profile, and turning his back to any volunteers on the engine who might recognise him.

Henry wonders if he's just noticed a jolt of excitement course through the accused. Had seeing the fire truck somehow stirred him? The interview is suspended and Sokaluk is moved away.

In the video's next sequence, he and the detectives are standing on the edge of the blackened eucalypt plantation. He's led them to an area metres from one of the points of origin identified by the fire scientists. This, he claims, is where he threw the paper napkin holding his cigarette ash.

'As far as I thought, it was dead and I chucked it out the window, but I didn't know it was lit up.'

The tree trunks are charred, the canopies the colour of rust.

'It was green at the time,' Sokaluk mumbles. 'Not green anymore. All burnt now.'

'It was green?' Henry asks.

'There's no more greenery and all the animals are gone.'

The detective tries what seems a neat shift. 'What are the weather conditions today?'

'Windy.'

Henry's tie is flapping over his shoulder; the air's rattle makes the conversation barely audible. 'What were the weather conditions like on Saturday?'

Sokaluk seems more competent in this interaction, quicker to respond. 'Don't remember it being windy,' he claims.

'What way is the wind blowing today?'

'Towards me.'

He's not looking at the detectives, he's gazing around at this new landscape. Standing there, Sokaluk appears curious. He would have seen some of the fire's damage when he retrieved his burnt-out car, but now, six days later, he actively takes in the ash-covered ground and the singed leaves pointing in the direction of the wind – the fire scene presenting its story.

Henry can't help trying to read Sokaluk's face, searching for some dawning sense of the violence that's been done.

Back at the police station, the detective clarifies the purpose of the field interview, and Sokaluk tries to slide away from his earlier admission.

Henry: 'You indicated the area where you lit the fire.'

Sokaluk: 'Well, I think I didn't really lit the fire.'

Henry: 'Sorry?'

Sokaluk: 'I didn't light the fire.'

Henry: 'Not, where you threw out the paper?'

Sokaluk, glancing over, as if sizing Henry up: 'Where I threw the paper out, roughly that area.' His finger is in his ear; he's stroking his neck, fidgeting more, moving his tongue in his mouth.

Henry: 'And that area was a tree plantation?'

Sokaluk: 'Yes.'

Henry: 'What sort of trees?'

Sokaluk: 'Eucalypts.'

Henry: 'And what was on the other side of the road?'

Sokaluk: 'Pine trees.'

Henry: 'So, you only saw one fire?'

Sokaluk: 'One fire.'

Henry: 'We've got information that there were two fires in that area.'

Sokaluk: 'Not by me.'

Henry: 'What can you tell me about that?'

Sokaluk: 'I only can say that there, where I was and stuff, not, I didn't light any fires. I think it was accidental.'

Henry: 'Yeah.'

Sokaluk: 'Now I'm paying for it.'

In this interview, his internet habits are again raised. Henry asks about the pornography. 'I'm referring to child stuff,' he says.

'I know that,' Sokaluk answers quickly. He sounds irritated, both with the police and himself.

Suddenly, he belches as if he's about to vomit. It's the most intense reaction he has had, and completely involuntary. In that moment, he seems to grasp that the trouble he's in is real.

Before this, while they were back recording the field interview, the suspect and Henry stood next to each other, both still for a moment, waiting for confirmation that the video was taping.

The detective, with knotted forehead and a biro in his top pocket, looks solemn. Where the camera might have picked up some undisguisable swagger around this arrest, the biggest of the policemen's careers, there's instead a sense of anticlimax, and a dawning horror at the pointlessness of this crime. The adrenaline, the urgency surrounding Sokaluk's arrest, slip away in a wash of futility.

Henry seems to be holding back, trying to suspend judgement. He'd dealt with murderers and rapists – men, usually, who often didn't seem very different to anyone else when you stripped away the layers. If he felt any prejudice towards them when he was gathering information, then he believed he was the wrong person to do the interview. In the days to follow, he'll oversee the use of cadaver dogs, and the sieving of crime scene areas for the bones of the dead. He'll send investigators to take DNA swabs from relatives who are trying not to imagine why they're needed. But right now, he is standing neutrally by the person who apparently caused this catastrophe, and it's impossible to tell if Brendan is excited or remorseful, nonplussed or uncomprehending.

Henry had been left raw by babies in cribs near his daughter's being given the last rites, and now he's standing on the edge of this crematorium of trees and animals and houses and people. And for years to come, this case will be the one that sticks with him. Every anniversary of the fire, he'll ask himself, *Why?*

Next to him, Brendan Sokaluk beholds the wreckage. He'd grown up around here. He'd fished for crays, and picked blackberries, and spent hours scouring this area and beyond like a modern-day ragpicker, an untouchable, for the junk other people left behind.

He knew these hills as well as anyone. An acquaintance from high school later recalled: 'Brendan could not read or write, but he had a real geographic talent . . . he had maps that he had drawn of Churchill town. The maps contained all the street names and the layout of the streets. It was as if he had copied the map, but he had done it himself. He even had deviations, of what he believed would be a better layout of streets. The talent pertained to the whole Latrobe Valley: back roads, fire access roads through the pine plantations and dirt tracks.'

Now, as he stands by one of these dirt tracks at the start of a blackened map, the damage from this fire stretches before him, over gullies and ridges in all directions. The hills are filled with exactly the kind of scrap metal – ruined farm equipment and endless strips of scorched corrugated-iron roofing – that he likes to collect. By lighting a fire, accidentally or not, he has turned the bush around his home into one vast junkyard.

In the background, the wind gusts in low keening cries. There's a sudden swell of ash-filled air and Sokaluk reaches up his hand to protect his eyes from the grit.

Part II

the lawyers

The afternoon of Friday 13 February, Selena McCrickard sat in a windowless office trying to concentrate. There was nothing on the walls and only a day's worth of legal briefs on her desk. Morwell Legal Aid was having staffing problems, and as she had family nearby she'd volunteered to spend a few weeks helping with the court lists. Just passing through, she'd found herself at the centre of a disaster zone.

Driving around the Valley over the years, she'd seen signs nailed to tree trunks: ARSONISTS DIE! Now many of those roads were sealed off as the fire kept burning, and no one knew when or where it would next spread. Images of ash, and victim photographs that you might see after a terrorist attack, filled every TV screen and newspaper. Arriving at work, she'd felt an eerie friction on the streets and wondered if the threat in those painted capital letters was now a real one.

The office manager came to her door. 'The police say they've arrested the arsonist.'

McCrickard, long blond hair around a striking face, looked back at her, not adding two and two.

'Come on. You're the barrister.'

It took less than ten minutes to walk to the police station, and McCrickard, a fit woman in her mid-thirties, usually went directly up the main street of Morehell, as she called the town with begrudging affection. She passed the FOR LEASE notices pasted on shopfronts, interspersed with charities and services for victims of crime, and then, closer to the court, lawyers' offices. Even as she moved she held herself straight, tensed, as if balancing on something.

In her spare time, the barrister was a surfer. The day of the fires, she'd tried to escape the ungodly heat down on the coast at Fairhaven, 125 kilometres the other side of Melbourne. In the water she found herself strangely buoyant. There seemed to be more salt than usual. She could taste it. She imagined she could see it. And when she emerged, her wetsuit was covered in a white crystalline lacework and McCrickard felt uneasy about the coming week. Later she'd remember this, because then there was no time to surf. And now, hearing of her new brief, she had the same feeling as when readying her nerves for a big wave. She'd hear its noise, feel the start of the rush – but just before that, and it would only be for a split second, there was the stillness you sensed before a storm hit.

This arson case wasn't what she'd had in mind when she'd volunteered to help out. McCrickard had just settled a gruelling local sex crime: incest, really young victims, the father pleading guilty. She was hoping for work that was less emotionally demanding. If she hesitated as a wave curled towards her, though, things always ended badly.

At the police station, she buzzed for access to the cells. No press were waiting here, but they were already outside the adjacent courthouse. She felt the hostility in the air. Inside the station, there were glances ending in head shakes; clenched, disapproving mouths.

The local police now regarded her with a new level of disdain. Some of them were still washing soot from their eyes and skin and hair. They'd spent the past six days standing at roadblocks, comforting the newly homeless. They'd guarded taped-off properties where the dead were sometimes their neighbours. And now here she was in her sleek business suit to protect the man they held responsible.

The police already had McCrickard's new client at the entrance to an underground tunnel that led to the courthouse, ready to put him in the dock. While he was being brought back to speak to her, she didn't need to sign in the way she usually did. The Arson Squad detectives wanted this handled quickly, and she was ushered straight through to an interview room.

She'd sat in this small, cold space before, talking through a glass partition to clients in various states of extremis; they'd be coming down off something, or panicking, or profoundly sad. Occasionally it was someone she'd come to know well. Perhaps this would be their final conversation before the Big House, the last hurrah, and the acoustics were such that if, behind them in the cells, another prisoner was wailing or banging, every noise was amplified and she could barely hear what her client had to say.

The man facing her had never been here before.

Sitting on the other side of the glass in his shabby green clothes, Brendan Sokaluk didn't exactly seem scared. He had a receding hairline partially disguised by its clipped style. Weight blurred his features – a doughy face narrowing to a soft chin and neck – but she could tell he wasn't young. Late thirties or early forties, perhaps, and wearing an expression she'd encountered many times before. Bewilderment: pure, lost-in-the-wildwoods confusion.

McCrickard took one look at him and thought, *Fuck*, this guy is a kid.

Through the tiny holes in the partition, they started talking; his speech was stilted and staccato, his gaze elsewhere. Over and over he repeated, 'I want to go home. I want to see my dog.' He seemed unable to grasp the fact that he was not going to be allowed to just leave and catch the bus back to Sheoke Grove. 'Who is looking after Brocky?' he kept asking.

Two things seemed clear to McCrickard: first, Brendan hadn't a clue about what was going on; second, there was no way she wanted him taken over to the courthouse for the filing hearing – 'kick off', his official entry into the justice system. She didn't even stop to explain the nuances of the Legal Aid application he had to sign, as she usually would: she did not have the time it would take him to comprehend it. She needed Brendan's signature and to then get back to the detectives and negotiate his non-appearance, with them and with the court. News of the arrest was still breaking. More media would be on the way to join those already in wait.

In the simplest terms, she tried to explain to Brendan that there was no chance of him getting bail, and right now it wasn't safe for him at home. How much he understood she couldn't tell. Probably bugger all, she estimated.

Outside the interview room, she went to find the detective in charge, Adam Shoesmith. He was already waiting at the courthouse. McCrickard left the station and walked around the corner.

The Magistrate's Court was another new concrete construction, opposite the town's rose garden. The juxtaposition amused her. Everytown's carefully pruned statement of order and civilisation was largely ignored by those who milled outside the building. Her clients commonly settled in to deal with their day's various legal matters accompanied by a passing parade of supporters and hangers-on, as were those seeking rulings against them. Here, in these ad hoc social gatherings, young kids pushed toy cars around defendants' feet, and all the clichés of the Latrobe Valley were on full display.

Some of McCrickard's colleagues at the bar rolled their eyes when they were given a brief in the Valley. For clients accused of rape this was the easiest place to get an acquittal. Selena could almost see the jury thinking, You got us into court for this? Often her clients were the fourth generation of – and she hated using the sanitised putdown – the 'welfare class'. It was normal for them not to go to school or work, as normal as the drug and alcohol problems, as Dad bashing Mum, as these inevitable days in the courthouse foyer where a shimmer of tension hovered like a heat haze, everyone braced for the next fight to break out.

The jokes about the region irritated her; the stigma of banjo country, of people being backward hillbillies living off the government. It was demeaning, the way it is when someone is rude about your family. An oversimplification of complex, interweaving factors. As a local you might be able to laugh at the trashiness, the easiest of targets – and the broader community certainly did – but McCrickard didn't extend the privilege to outsiders.

Today, however, the courthouse was different. There were no kids strapped into strollers as if for a visit to the zoo. Today the atmosphere was laced with hate. The waiting press and onlookers could have passed for the first sign-ups to a lynch mob.

Inside, past the security screening, a narrow corridor led to court number one. It was full of reporters, who went silent and did not move to let her pass. Vicious bitches packed with make-up, she thought. They were right in her face – and the looks they gave her! It was as though *she'd* lit the fire.

Brendan was waiting in the dock. The press stood around eyeing this prize monster, who sat expressionless, staring straight ahead.

McCrickard started negotiating his non-appearance with Adam Shoesmith. The detective agreed they had to get Brendan out of town as fast as possible. He'd been the one to ring Legal

Aid and ask for the duty lawyer to come and represent the accused at the hearing. It was up to the court to give permission for him to stay away. But there was nothing to be gained by having this guy on show. The hearing was merely procedural, and a chance for the defence to request service of the hand-up brief – the document that would include Brendan's record of interview and the prosecution's witness statements.

When Brendan had been removed, and his matter was finally called, McCrickard tried to apply for a suppression order. She was on her feet arguing the case as a series of faxes were placed in front of her. News agencies from across the country wanted this, wanted that, wanted her to stop and wait for them and would slap down an injunction if she didn't. Working with this level of raw aggression was a test of nerves.

The magistrate declared a suppression order unnecessary. 'We don't need to mention any names, because this is just a filing hearing, so we're just talking about dates.'

He set about adjourning the matter to a date and court in Melbourne.

'Thank you, Ms McCrickard, you're excused.'

In the crush of the courtroom, there were two other people supporting Brendan. Kaz Sokaluk, a reticent man in his late fifties, was already somewhat hidden under a long ponytail, beard and glasses. He had another son with him, and they looked equally overwhelmed. Kaz was not used to attention of any kind, but just now his house was under siege from the media and the phone would not stop ringing. His wife was too distressed to come out. Neither of them could believe their son capable of the things he'd been accused of doing. Brendan had been picked on all his life, for being slow and unusual, and his arrest felt no different.

McCrickard sought to reassure them both but there was no point talking seriously with dictaphones shoved right in their faces.

'I'll have to call you later,' she told the pair. 'Just get out of here. There's way too much heat on. Go home and I'll speak to you soon, alright?'

The court staff managed to sneak the Sokaluks out of the building.

The barrister saw her client again briefly. He was sitting still behind the glass of another small interview room, just off the underground tunnel, unmoved by the drama surrounding him. He didn't seem to recognise her. Once more she had the feeling that she was dealing with a child.

McCrickard was well aware of the long history of intellectually disabled people admitting to crimes they'd never committed. Brendan waiving his right to a lawyer during the police interview was typical of someone unable to grasp the gravity of their legal situation, or the consequences of a confession. People with an intellectual disability were more vulnerable to all the tactics of ingratiation and threat that police commonly used. They were more likely to confess, wanting to give the right answer to please authority figures, especially those who appeared friendly. Even more so if the interview had been lengthy and they'd been provided with no assistance – or only a barely trained independent third person. The man in front of her may have thought the police were right about circumstances he didn't entirely recall, particularly if the accusation dovetailed with a general sense of shame. Thrilled and terrified by the detectives' attention, Brendan may have claimed responsibility for the fire in a misguided attempt to big-note himself.

Reintroducing herself, she was blunt. 'Really, Brendan, the game plan today is to get you out of here. We've got to get you out of Morwell.'

She had a gift for talking to people from behind glass, building some rapport quickly. 'I know you don't understand now,' she told him, 'but you're going to have to trust me. I'm going to look after you, but you have to trust me, alright?' McCrickard had a smoker's stripped vocal register that gave her words a raw authority. 'I don't want you to talk to anyone.' She repeated herself slowly, stressing each syllable. '*Don't talk to anyone, anymore.*'

Brendan was led away and the barrister walked back from the tunnel to the court's foyer. The din! As though something long-buried underground had emerged, hissing. The plate-glass windows had morphed into a wall of waiting eyes. There's no fucking way I'm going out there, she thought. The press had multiplied, their lenses and microphones all emblems of hate. The usual atmosphere of the place, the everyday hopelessness combined with tension that seemed to have a scent, was overpowered by loathing.

The court staff tried various ways to smuggle her out of the building. Just as they resorted to walking her through the tunnel back to the police station, the media dissipated. They'd got wind that Brendan was being loaded into a prison van. As the mob raced over to get pictures, McCrickard made her escape.

Behind the reporters, Hazelwood's eight chimney towers went on making their cloud forest: life as it was known.

The next day, the front page of *The Age* showed a man hanging off the van, pounding on the reinforced windows. The court had printed Sokaluk's name on their daily list, and the media had connected the dots: arson causing death, and knowingly possessing child pornography. Hate sites devoted to 'naming and shaming' the accused were now all over the internet. Soon a website titled 'KillBrendanSokaluk' appeared, along with a 'Brendan Sokaluk Must Burn In Hell' Facebook page, and another group purportedly offering ten thousand dollars to anyone who killed him. The mainstream newspapers published some of the posts:

'Let me hurt him, burn him, put a bullet and a knife in every orifice of his body.'

'He should be thrown into his creation and left to burn.'

'Tie the bastard to a post and put a ring of fire around him . . . let the fire make its way to him and make him suffer like the other 100's of people had to endure.'

Selena McCrickard read these offerings with increasing dismay. She knew virtually nothing about her client, and was now encountering the most lurid, distorted version of him.

Some of the coverage was like a pastiche of small-town gossip. In the *Herald Sun*'s tabloid exposé, 'Australia's Most Hated', it was reported:

Sokaluk was an enigma in a town where he grew up [but now] 'Everywhere you go people are talking about it, even at the shops,' one resident said . . . One neighbour, who asked not to be named, said Sokaluk played bingo [but] never mingled.

Residents had become angry over the past year as Sokaluk's collection of car parts and refrigerators grew and littered his driveway . . . '[A]ll the other houses around here are pretty nice looking, but that one made the street look untidy,' one neighbour said.

In another *Herald Sun* article, titled 'Secret Life of Arson Accused', she read: 'Accused arsonist Brendan Sokaluk lists candlelight, thunderstorms and skinny dipping as turn-ons on his social networking website . . . A loner who went to a special school, he tried to become a CFA volunteer for years but was always knocked back . . . One [of his social media posts] written on November 26 said: "We are compelled to do what is forbidden."' Surely, McCrickard thought, this was the tag line of a DVD, rather than a sentence the man she'd met had composed.

The story was picked up internationally. Online and in the UK, *The Guardian* reported: 'The man charged over his role in Australia's deadly bushfires was a loner' – being isolated, she noticed, was itself nearly a crime – 'obsessed with fire and was bitter about an ex-girlfriend.'

A photo of Brendan's ex-girlfriend, Alexandra, a pretty peroxide-blonde wearing a bright yellow CFA uniform and angel wings, had exploded over the internet. It was her profile picture on the internet forum Myspace, where, newspapers reported, she claimed that as a child she had been severely burned and spent a year recuperating on a burns ward. She also wrote of her dog, named Miss Pyro.

Her image was paired with an unsmiling self-portrait of Brendan. Wearing a black beanie, he'd photographed himself in his dark bathroom, sunlight from a narrow window reflecting in the mirror like a white blaze.

The couple had evidently had an unhappy break-up. Newspapers quoted Brendan's Myspace profile: 'My interest are to enjoy life to the fallest and not with Alexandra because she roots behind your back and lies a lot. I'd like to meet my sole mate and not some old hag.'

McCrickard started calling the police, demanding they act to have these images and websites taken down, but even as she argued Brendan's case she felt it was too late. The lynching party was electronic: how could he not be stigmatised by this media coverage? With his image intercut with those of the fire, would there be twelve people in the state who could sit unbiased in a jury box?

Before long she was also butting heads with Detective Shoesmith about the child pornography charges. In crime parlance, they looked to her like a 'burger with the lot', similar, for example, to an armed burglar also being accused of double parking and firearm possession. She felt the Arson Squad had charged Brendan too hastily and caused immeasurable reputational damage. Paedophiles and arsonists were the pariahs of modern Australian life – to be both rendered someone the ultimate outcast.

The barrister knew all the clichés about fire-setters also being sexual perverts. Sex and fire was an old, old fusion. Sigmund Freud had not been required reading in any law school course on fire-setting – and Legal Aid did not go in much for psychoanalysis – but in *Civilization and Its Discontents* (1930), Freud synthesised the idea by writing that fire-lighting was a regressive attempt to master the threats and uncertainties of the natural world: 'In man's struggle to gain power over the tyranny of nature, his acquisition of power over fire was most important. It is as if primitive man had had the impulse when he came in contact with fire, to gratify an infantile pleasure in respect of it and put it out with a stream of urine . . . Analytic findings testify to the close connection between the ideas of ambition, fire and urethral eroticism.'

A few years later, Freud provided his further thoughts: 'The warmth radiated by fire evokes the same kind of glow as accompanies the state of sexual excitation, and the form and motion of the flame suggest the phallus in action.'

Eighty years on, fire-lighting was still widely considered to deliver an erotic thrill. Back at the Morwell police station, some local detectives who had been inside a lot of fire-setters' houses reckoned they'd found uncommon amounts of sexual paraphernalia. And some of McCrickard's colleagues, criminal barristers who went on to defend Sokaluk, privately also believed in the connection. One had defended a man who would lie down by his car, nearby his blaze, and masturbate; another defended two intellectually disabled men who would head into the bush, light fires and jerk each other off; and yet another defended a man whose calling card was setting fire to women's shoes. Despite studies that found no proof of any link between sex and arson, in Sokaluk the two were now related.

Adam Shoesmith had never worked in any police department dealing with child pornography. He'd never been exposed to the kind of images the forensics team pulled off Sokaluk's computer, and for him the material was genuinely shocking. But McCrickard's point was that these charges could have waited; Brendan wasn't about to get bail. Arson causing death was essentially a homicide offence, and he also faced a multitude of ancillary indictments relating to the fire. The man was in terrible trouble. In comparison, the child porn charge had complications. There were unresolved legal technicalities regarding whether the images were, as Sokaluk claimed, 'cookies' that had popped up when he was looking at adult porn. By the time the case reached committal, these charges, regarded by the prosecutor as necessary but ultimately a distraction, had been carved off along with the other 180 counts of recklessly causing serious injury and damage.

In her most frustrated moments, however, McCrickard couldn't help feeling that the porn charge had been thrown in to make the police look like heroes and her client as nasty as possible. She knew her perspective was different. She hadn't been present at the blackened crime scene as it was photographed, and she hadn't had to knock on doors and tell people what happened to their relatives.

Nor did she take the possession of child pornography lightly. Once, reading a judgement in a case involving a man downloading live child-sex shows, she'd found a sentence about the abused children screaming for their mothers. She had reached the bathroom just in time to vomit. One day, years later, her own child called to her after she dropped him at child-care and out of nowhere that line reared up in her mind, and there was no way to explain to a four-year-old why suddenly she felt so undone.

Selena McCrickard was uniquely placed to separate the offence from the offender, the accused from the accusation. As a newborn she had been taken home from the maternity ward to Hayes Prison Farm, a low-security jail outside Hobart where her father worked as the superintendent. She grew up having the run of a farm with beef cattle, dairy cattle, a piggery, chickens and a vegetable garden. The produce fed the prisoners and her family, with the surplus donated to the local community.

For the most part, she was aware of what the different inmates had done. Many of them were open with her about their crimes. Aged five or six, she'd pad around in the school holidays after one prisoner, Shane, a regular who tended the huge lawns and rose beds of the superintendent's rambling weatherboard house.

'Shane's back,' her father would announce periodically, to her mother's delight: the mowing would now get done. Shane himself seemed relieved to have a break from the booze and the drugs, and just take it easy in the garden, chatting with this little girl about his latest escapade – getting loaded and stealing a car, say. Often she heard quite detailed descriptions of the destruction these men had wrought in their lives outside, with them confessing, 'Yeah, I shouldn't have done that.'

The deal was that anyone who was near their house was a low-risk offender, and because security was minimal, for their own safety not many child molesters were sent there. Selena never felt at risk: to her this was just the way it was. Those in for murder, she noticed, often wangled the best jobs because they were, literally, stayers. One man, Paul, who was sentenced to twenty years, had experience milking cows and so the prison dairy had never run better. Paul used to piggyback Selena around, this gentle-seeming man who'd killed his girlfriend. Good and evil were far from categorical concepts, and liable to morph from one moment to the next.

Her childhood trained her for all the people she'd meet later as a Legal Aid barrister. Covering her desk on any given day were cases involving defendants with congenital or acquired brain damage: people with fetal alcohol syndrome, long-term alcoholics or drug users, those who'd been cognitively impaired in accidents, fights, or by various forms of abuse. Selena McCrickard calculated that more than half her clients in the Latrobe Valley courthouse had some form of intellectual disability. And sometimes she remembered Richie, a mentally impaired old man who would be released from Hayes Prison Farm then go and smash some windows to be allowed back in. Thirty years later, a lack of understanding, resources and support meant that people like him were still cast into the criminal justice system, and having no voice they depended on someone standing up for them in court.

In the attitude of some of the police, she recognised the familiar question: How can you sleep at night? Or more politely: How can you defend him? Her unspoken answer was: Well, because I'm a lawyer, not society's moral guardian. And so, on a Saturday, a little over a week after Brendan's arrest, she went to visit him in the Melbourne Assessment Prison, a red-brick fortress in the heart of the city. She was led up a set of stairs to an area of the prison she'd never been, and taken to a bland room with a table and chairs.

'Normally we'd never bring you here,' a senior warden said, explaining that it was too dangerous to place Sokaluk near any other prisoners. He was in lockdown for his own protection.

Brendan was escorted into the room wearing a prison-issue tracksuit. There was that odd, clumsy gait, and when Selena caught his face there was something in his expression that didn't look quite right. This was the animal moment when a defence barrister almost unconsciously sizes up a client, wondering how he or she will appear to others in the artificial jungle of the courtroom.

From the warrant and hand-up brief, Selena knew the police didn't have much evidence on Brendan. But now it struck her again that anyone seeing him near the fire would likely have made all kinds of prejudiced assumptions. She suspected he wasn't much used to candlelight or skinny-dipping, although nor did he present as 'bitter' or 'an enigma'. What *was* clear was that this man had 'issues', and people like him often did stuff, often unexpected, off-the-wall stuff such as driving around the countryside on a 47-degree day searching for scrap metal. It didn't mean he'd lit a fire. She knew he'd been an early caller to 000, reporting the blaze, but so what? Plenty of people were calling the fire through. Maybe he'd found himself in the midst of the drama and wanted to be a part of it, then, being the misfit on the scene – as he probably was everywhere – all fingers had pointed his way.

Years earlier, Selena had defended a man accused of lighting fires around the Macedon Ranges in central Victoria – a strapping, handsome, bloke-next-door type who was acquitted. Brendan was at risk of bias in a way the attractive accused man was not.

Again she was unsure if Brendan recognised her. She told him she was here to check that he was okay.

Brendan had a sore back from lying down so much in his cell. He was trying not to make any noise, hoping the other prisoners wouldn't know he was there.

Selena explained that one of her colleagues at Legal Aid had handled his suppression order – no one was allowed to name him or show his photograph now. On the cell's tiny television he'd seen the pictures of himself and his ex-girlfriend and his plain brick house. He was worried for his next-door neighbours. What if someone blew up his place and they were injured or killed? Perhaps it would be bombed by the mean man he met late the night of the fires who told him he'd kill whoever was responsible . . .

That was how it was talking to Brendan: little scraps of fantasy, of make-believe; ideas from whatever cartoon he'd just watched seemed to float into the discussion. His attention span was very short, he had to be brought back round to the point:

yes, he'd noticed that the footage showing his photograph had suddenly stopped. But Selena seriously doubted how much he comprehended.

In the months to come, it struck the barrister that Brendan's interest in his legal affairs extended only so far as wanting them to end so he could go home. His conversation revolved around a limited range of topics that interested him. She had a V8 Holden ute, useful for carrying her surfboard, and he liked chatting about the engine, the tyres, how it ran. Otherwise, his areas were children's television, fast food – he'd suddenly start rhapsodising about KFC – or people being mean to him. Talking with him involved a series of strange, self-absorbed non sequiturs.

This was an ongoing shock: his seeming ineptitude and the fire's vast destruction. Brendan maintained he hadn't lit the blaze, although he showed no curiosity about its impact on the community in which he'd grown up. And when they ventured too near this topic, his manner let it be known that it was all a bit too hard for him. The reasons why it was too hard weren't entirely clear. Sometimes Brendan would shut down, barely seem able to respond.

It is common for people accused of crimes to be overwhelmed, as if by a nightmare. McCrickard had seen clients break down, crying, 'I'm not guilty! Why am I here!' Or conversely, 'Yes, I did it, and I feel like killing myself!' The level of emotion often reached such extremes she would leave these interviews exhausted. Sometimes it was straight desperation: someone about to hit crack withdrawals claiming they needed bail to care for a sick grandmother. (Last night, when you were robbing the servo, Selena would wonder, who was looking after Granny then?) But it was an understandable human reaction. She could read it. Brendan's expression and speech, though, were unchanging, his gaze elsewhere.

Selena was not encountering the same person the police were. The Arson Squad detectives had decided his vagueness was an act, a claim the barrister found ridiculous. To them, his apparent lack of remorse verged on sociopathy. But Brendan's aloofness, his almost mechanical egocentrism, was what made her suspect there was more going on for him than an intellectual disability. Although on the night of his arrest a forensic doctor had found him fit to be interviewed, this was not the same as being fit to stand trial. Selena had major apprehensions about her client's capacity.

She wanted him to have a psychological assessment. She'd thought this the moment she met him. She thought it again on her first visit with him in the Melbourne Assessment Prison.

She was sitting opposite a man stooped by habit or fear, who seemed impervious to the devastation he may have wrought in the place where he was raised. The only thing he'd wanted his family to bring him in jail was a photo of Brocky. Selena tried to tell him that his pet was alright, that it was being cared for by neighbours, a family with children, but Churchill wasn't safe for Brendan himself. And all the time she was thinking, Is someone going to try to kill you in here?

Brendan later told his parents and lawyers that other prisoners had twice tried to murder him while he was on remand. He claimed that after he complained about corrupt guards bringing in KFC for select prisoners, they left paperwork relating to his child pornography charges where other inmates could see it, a greater black mark apparently than the arson charge. His parents worried he was too naive to understand when to keep his mouth shut.

Most prisoners McCrickard dealt with either already knew the ropes when they got to jail or quickly learned them, but this guy had no idea. He was watching *Thomas the Tank Engine* with the sound down low and trying to pass the time doing a puzzle book. His soft, plump body, his wooden movements, his fidgeting made him seem more vulnerable. He looked lost, and she found herself reaching into her wallet and handing him fifty dollars. Now he could at least buy a packet of cigarettes or make a phone call.

He took the money, but all the time kept asking the same question, and she kept answering it the same way: 'You can't go home at the moment, Brendan. It's not safe for you to go home. There are a *lot* of people who are really, really angry at you. They think you've done something very bad. Even if we get you bail, you may *never* be able to go home.'

In the weeks, then months, following Brendan's arrest, Selena McCrickard had to drive to work past the fire damage. Slowly, the blackness on either side of the road began to change. Tiny shoots pushed their way through the charred ground. Tree ferns that had been burnt to stumps unfurled new green fronds, the showy ones at the funeral. Plants that needed fire to trigger a grand seed release or to crack their seeds open now had ash-rich and fire-warmed soil to stimulate their growth, and the forest's cleared canopies meant there was no competition for light.

On different days the vivid shoots could look miraculous or pitiful, depending on her mood. They could look like a flashback:

an epicormic bud that may have been lying dormant under the tough bark of a eucalypt for years, decades even, would sprout bright red leaves, as if in budding these little tongues of flame the trees were acting out what had happened, the forest also unable to forget the fire.

With a rape or an assault case, the damage was hidden or close to buried, denied. Here, it was everywhere. She'd turn a bend in the road and glimpse another crime scene. The trees brazen co-conspirators, because of course the eucalypts were made for this crime of fire.

In the weeks after Black Saturday, animal shelters around the state weren't treating as many creatures as they'd expected. The burning bush had been too severe for most to escape. Selena had a series of pets from rescue homes – she'd even offered to care for Brendan's dog – and details of dead wildlife and animals were among the saddest sections of the police witness statements she read. The RSPCA calculated that one million animals were killed in the fires or died afterwards, from starvation or as easy prey. The Victorian Association of Forest Industries estimated this figure to be in the millions. Ash in streams decimated aquatic life, and surviving animals on the ground would have to endure the winter until spring provided cover and food. If it did. There were areas where fire had killed everything, burning through the surface of the earth and cooking the layers of sediment, like clay in a kiln. It could take decades, possibly longer, for certain species to return.

Some of the places she drove past had no regrowth, just a sooty palette of greys. Then, to prove the fire's inconstancy, the road would curve and she'd find new shoots clothing the trunks and branches of eucalypts in moss-like buds, and later a profusion of bright stems. Within a few months of Black Saturday, the trees resembled maypoles, wrapped in curling red and green ribbons. The forest a kind of pageant.

Selena would then turn into the wide, empty streets of Morwell, park outside Legal Aid, and brace herself for a day of more fire.

To facilitate Brendan's assessment by a forensic psychologist, she needed to provide her client's basic history. Brendan could not easily give her a coherent outline of his childhood. His cognitive impairment was no doubt part of the problem, although she began to suspect he also didn't *want* to remember it.

'I'd like to talk to your mum and dad,' she had eventually told him. 'I need to ask about when you were a kid, and how you were at school, because it's going to be important later.' He gave his permission.

Selena knew that Brendan Sokaluk's parents, Kaz and Lou, had barely been leaving the house. It was the same modest brick place they'd lived in for nearly forty years, but in the days after their son's arrest, reporters waited out the front. Cameramen followed anyone coming or going, and long after the cameras left, the Sokaluks felt they were under surveillance. Even the simplest social transaction became excruciating. To send a letter at the post office meant readying themselves for the gaze of the whole community, including the postmaster, whose house had burnt down. Any unknown mail they opened gingerly. A death threat had been sent to them via the *Herald Sun,* prompting the local police to ask if they'd consider getting out of town for a while.

No, they'd answered: if they ran, they might have to keep running. Better to stay, with the curtains drawn, and hope there'd been a mistake.

Selena would later read, in the finalised police brief of evidence, that when the Arson Squad detectives knocked on the door and told Kaz his son had been arrested for lighting the fires, he immediately replied, 'No, I don't think he'd do that.'

The detectives were in the living room when Lou came home. They told her why they were there, and noted that she turned to her husband, saying, 'I don't really think he would, would he?' Even as Lou spoke, the barrister imagined, life as she knew it slid out from under her.

When Selena called to introduce herself, Lou asked warily, 'How do I know it's not someone pretending to be you?' She also wondered if her phone was bugged.

Selena said that Brendan's calls from the prison were most likely being recorded, and indeed, later some were transcribed and tendered as evidence.

On his son's first call home, Kaz, after checking Brendan was okay, was blunt:

Kazimir: 'Did you do it or not?'

Brendan: 'I don't think I did.'

Kazimir: 'What do you mean, you don't think?'

Brendan: 'I think it's a bit more accidental . . . plus those two coppers, they said if I helped them they would let me go home. So, I reckon I was set up probably . . . they just attacked and bullied sort of thing and I tried to explain . . . it's an accident . . . I can't burn a forest down, I love it too much . . .'

Everything had happened so quickly. When the Sokaluks realised that Brendan was at the police station, Kaz had gone around there and pleaded for his son to be given a lawyer. He'd told the coppers Brendan wouldn't be capable of knowing he needed one, even if one was offered. Kaz believed he'd been fobbed off: the police just looked him over and calculated they could ignore him. This felt like a legal mugging, as though his son had suddenly been snatched – and, worse, as though the authorities had needed someone to blame, fast, and Brendan was an easy target. Everyone knew that Brendan usually spoke 'rubbish', 'a load of rot': what if he had confessed to dropping a cigarette butt to get out of the watch-house cells and back to his little dog?

When Kaz and Lou Sokaluk arrived in the interview room of Morwell Legal Aid – a space with frosted-glass walls, industrial furniture and carpet in a vivid shade – Selena knew it was a big deal for Lou to have even left the house.

Brendan's mother was in her late fifties. She had been athletic in her youth and remained a strong woman, with a kind but plain-spoken manner. Her son's features were visible in her round face, but she would light up and reveal anger or distress or suspicion – and later her humour. It took a while for people to see Lou's softer side. But Brendan had told his mother that the lawyer had opened up her wallet in the assessment prison and given him fifty dollars, and Selena guessed this wasn't a small amount of money to Lou. Kaz had long been unemployed, and Lou worked on a dairy farm, milking. She tried to pay the money back but Selena wouldn't take it. The barrister got the feeling that not many people had shown this woman's son much kindness.

McCrickard asked Lou to dig out Brendan's old school reports and any medical records she could find. She also needed an account of Brendan's life, and so the couple tried to tell his story.

Louise and Kazimir Sokaluk had moved to the Latrobe Valley early in the 1970s, hoping for a better future for their children. The government-owned coal industry looked like it would provide the security to raise their toddler, Brendan, and Jamie, his younger brother. Kaz found a job at the Hazelwood Power Station as a trimmer. It was dirty work, using shovels or rakes to keep the coal level on the conveyer belts, and cleaning up spills off the belt or dredger. But Kaz had known hardship. He'd immigrated from Poland as a small child, his family postwar refugees who settled near Lou's home town north of Melbourne. The pay at Hazelwood was good enough for them to buy their own house, at a rate subsidised by the State Electricity Commission.

In those days, Churchill had a population of 2500. People were moving in from the nearby town of Yallourn, which was being razed for the coal underneath. In Yallourn, in winter, a fog full of coal dust would lift at 11 am then descend again a few hours later. In summer, you couldn't take your baby out in the pram for some sun: the brown particles fell on everything outside, and settled inside on food and furniture.

In Churchill, though, the air was clean and the neighbour-hood was full of other young families with small children. You didn't have to lock your doors. Kids played outside and only returned when the streetlights came on.

The Sokaluks had a third son. Money was tight, but it was the same for everyone. Around the street, those with television sets would move them to their front window and their neighbours would join them in the garden to watch programs together. Local women steadfastly baked and fundraised for a swimming pool, but, lore had it, their money went to the production of the Big Cigar. So instead, people took their kids swimming in the power station's pondage, a 520-hectare, man-made lake filled with coolant water. Waste heat kept the pondage at a bath-like forty degrees.

By the time Brendan reached kindergarten age, his mother knew that he was different to other kids. He started school with limited speech and was unable to control his bowels. Doctors suspected he was brain-damaged after a long and difficult labour. Lou, who was then working full-time in a Morwell factory making electrical equipment, took her son regularly to speech therapy. She wondered if it helped. At school, unable to easily communicate, Brendan found it hard to learn to read or write. Other children scorned the boy. He grew to prefer his own company.

His mother tried to find ways to include him in the town's life. She took him to soccer lessons, where before long he was surrounded by other children who were all kicking him. No adult intervened. It was as though no one knew what do and so, in their awkwardness, did nothing. Lou took Brendan back home, but with limited activities available for kids in the town, they returned the next week. Again the other children set upon him, and again none of the other parents – her neighbours and acquaintances – moved to stop them.

This new community had no old people, no history, and no long-running feuds, but also no deep-rooted connections. In such a small town, everyone knew everyone else, and everyone knew Brendan had difficulties. But people had come here to make a new beginning, and this boy was not what they'd pictured. Did they feel that his strangeness reflected badly on them? Or that if they got too close it could be catching? Lou had always got along well with the disabled people she'd known: Kaz had a younger brother with Down syndrome whom she adored. In a mental black book she silently recorded the attitudes of otherwise good people, many of whom she considered friends.

Brendan never played sport again, nor joined any other club. In his presence, people were casually cruel: 'vegie' or 'retard' or 'spastic' slid easily off their tongues. He had no friends, and a mania for sameness. If Lou moved his things when she dusted his room, she'd return to find them rearranged exactly the way they'd been. Once, when a substitute teacher took his primary school class, Brendan hid in a cupboard and had a frenzied tantrum when the other teachers tried to drag him back to the classroom.

But he did love to draw. He drew plans of buildings and, panning out further, maps of the area where they lived. In these pictures there was an uncanny aerial perspective. Brendan had never been on a plane (even at the time of his arrest he'd never flown anywhere), but he would draw Churchill from above, making adjustments where he thought necessary. He redesigned the town – and thus his life – in the only way he could. This stolid, clumsy child floating high above his tormenters. The forest and the roads and the buildings all free of people.

There was one boy, Lou told Selena, perhaps five years older than Brendan, who often appeared by the side of the road with his bike, which had a tall, orange safety flag attached. He was the only other noticeably disabled child in Churchill. He'd stand waving at passing cars. Most people ignored him but Lou always made sure she returned his wave. Her house backed onto the local police station, and one time she heard a policeman giving a belting to teenagers who'd been bullying the boy. Part of her wished someone would do that for her son.

When Brendan reached high school age he attended Morwell Technical College, a fifteen-minute bus ride away. His mother felt he'd do better in life if he went through the regular system rather than to a special school. But Morwell Tech was as tough as schools come, in a town where, depending on the wind, some days a ripe sulphuric smell descended, not from the power station, but the nearby paper mill. Depending on other factors, Brendan would routinely be set upon.

One day, he got off the school bus with faeces smeared on his back, courtesy of a student. Another day, he returned home with globs of mucous that his tormentors had coughed up and spat into his hair. He still had trouble controlling his bowels, and if he soiled himself on the school bus he was treated like *he* was the shit. A subhuman with a surname easily turned to *Suck-a-lot* or *Suck-a-cock*. The codes of teenagers, and their capacity for cruelty, were incomprehensible to him. Each bus ride offered a unique form of torture, until his parents pulled him out of school at the end of year II. He had passed nothing. He could barely read or write.

Lou helped her son find a few menial jobs. He was always laid off, but in 1988 his luck seemed to change. Through a special program for the disabled he managed to gain employment as an assistant gardener at the Gippsland Institute of Advanced Education, soon to be amalgamated with the prestigious Monash University.

The institute sat awkwardly within Churchill. Very few of the academics lived in the town, and if any of the snobbier lecturers' children attended as undergraduates it was considered mildly embarrassing. The campus had been founded in the 1920s as a technical school to train SEC workers for the local power stations, and later the faculties were expanded to remedy the high dropout rate for Gippsland students at campuses in Melbourne. An engineering department was established because, the SEC found, the wives of electrical engineers needed in the power stations didn't want to come and live in the Valley. By training local engineers, the rationale went, they'd marry local girls and stay.

For academics, the Gippsland campus lacked prestige but was a pleasant, low-pressure place to work. On the far reaches of the empire as it was, no one high up saw or particularly cared what was going on. Most students were the first in their families to attend university, and if any showed signs of excellence they tended to be funnelled back to the campus in the city. Many of the administrators were ex-SEC, and the commission's informal motto of 'Slow, Easy and Comfortable' applied to this work environment too. There was a golf course within the grounds. There was also a culture of graft and corruption. Once, the university took an inventory and found that one-third of their computers had been taken. Everyone was in the habit of looking the other way, enjoying the vista of sweeping, immaculately kept lawns.

Lou Sokaluk believed Brendan had arrived at his gardening job relatively innocent. But in this town everyone knew your rank in the hierarchy, and he was now spending his days with the siblings and friends of the kids who'd tormented him throughout school – people often with their own issues, who had no training in working alongside someone with special needs. (Selena reckoned that having a disability in the Valley, especially in the 1980s and '90s, would have just given local folk another reason to pick on you. Brendan was the butt of jokes amongst people who were themselves the butt of jokes: when you feel downtrodden, she supposed, it's a relief to find someone else to tread on.)

Here was a backward eighteen-year-old who still suffered from incontinence. Perhaps thinking he was acting in the spirit of his place of work, he tried to engage his colleagues with boastful stories and practical jokes. He would hide in bushes then jump out when a co-worker came looking for him; he would sneak up behind other gardeners and tap them on the shoulder, chortling – the stunts of a kindergarten child. The gardening staff almost uniformly resented working with someone so irritating, and who needed near-constant supervision.

Years went by and Brendan's parents helped organise him to buy a house not far from them. Selena could see in the police photographs where his mother had set about furnishing it, adding doilies and knick-knacks, some pictures on the walls. Lou gave him a weekly allowance from his wages (otherwise Brendan would spend it all at once) and used what was left to pay his bills and mortgage. For a while – long before Alexandra – Brendan had another live-in girlfriend and he'd installed ramps to the door for her wheelchair, but this relationship didn't last either.

The region to which Lou and Kaz had relocated to give their sons a better chance must have been rapidly changing. By the mid-1990s, Kaz had already been pensioned off for a decade with his bad back, and the SEC was about to be privatised. When a conservative Victorian government split up and sold the state-owned power assets, over 7500 people lost their jobs in the Valley, and the 23 billion dollars in sales profit flowed elsewhere.

Instantly, the region became a different kind of place. There had always been poverty hidden in the hills, but now the towns were also full of people on benefits. Those who stayed watched a rust belt form in real time. The Valley became a human sink, a place people ended up.

Sometimes, as Selena drove to work and saw Hazelwood in the distance, Bruce Springsteen's 'Youngstown' came to mind. He sang of smokestacks stretching into a dirty sky like the arms of a deity, the rise of industry and the looming fall. Along with the song's narrator a lot of the people Selena met felt they had been forgotten. They lived beyond the sight of those with influence, amidst the symbols of the unloved past.

The eight chimneystacks of Hazelwood were now *the* iconic image of the country's dysfunctional climate policy and an emblem for the campaign against global warming. The power station was the dirtiest in the OECD, reportedly pumping 16 million tonnes of carbon dioxide into the skies each year, although the figure was probably conservative. It turned out the total emissions from Hazelwood Power Station weren't actually measured. Once every six months the discharge was recorded from one stack at a time, but that was it. No one really knew how much particulate matter floated around, and in the meantime the state government's Environment Protection Authority sold the company owning the mine, International Power, emergency approvals to pollute further, including into the Morwell River.

It wasn't surprising that many locals were distrustful of authority. Why put your faith in the power companies that slashed jobs and safety standards? Or in politicians – even the ones championing coal – who didn't want to talk about the local kids with asthma and other respiratory problems, or about the adults here, who lost more years to disease than in any other region in Victoria. And no one saw environmentalists crying themselves to sleep over those who'd be left behind if Hazelwood or the other power stations ever closed.

The Latrobe Valley's brown coal deposits, the biggest in the world, stretched for sixty-four kilometres in one direction, fifteen kilometres the other, with a depth of 180 metres. The towns clung to the coal. Morwell sat on the very edge of a mind-bending pit, and to Selena it could feel like it was close to falling in. The highway and the Morwell River had both been diverted to extend the mine's lifespan. And older people were reminded of when Yallourn's Olympic-size swimming pool, cricket and football grounds, cinema, croquet lawns, avenues of trees, golf course, schools and library, not to mention houses, were all demolished to get to the coal underneath. Geology ruled the Valley. And it was as if the volatility of the coal affected those who lived alongside it. Brown coal is unstable – two-thirds water when pulled out of the ground and highly combustible. People's friends and family worked cutting the stuff out, burning it, and then everyone breathed in the vapours of strife.

Soon those bare walls in McCrickard's office were covered in butcher's paper. Mapping out the brief of evidence was the only way she could get her head around the multitude of witnesses who had seen Brendan on Black Saturday, or had some tale to tell about his past.

This brief was coming to her in stages. The police were waiting on various laboratory tests, and some people weren't yet ready to be interviewed, while others needed to be interviewed again. Nevertheless, the document was already vast, the index of witness statements resembling a phonebook.

Selena spent long hours, pen in hand, reading meticulously through each statement, then rereading for what she may have missed. Arson is a difficult crime to prove and it's the granular detail that convicts or acquits most people. Text messages, phone calls, time-stamped security footage. Small things could be vital, and she couldn't skim over any evidence, any of the countless incidental details of horror.

A man who had a resident koala on his block found it dead at the base of its favourite tree. A woman discovered unburnt magpies all over her property, their wings still outstretched as if they'd just dropped from an airless sky. Someone had lost letters that were the only link to a loved one who was gone. The war medals, or model aeroplane collection, or essential work tools, or family photographs – everything tangible and intangible that a house, a shed, a farm, a *life*, contains – all of it had burned.

And yet, as more information came in, two main possibilities were opening up for Brendan's defence.

The first was timing: witnesses had wildly different recollections about the fire's beginning. As was common in bushfires, people were in their houses with the air conditioner cranked up, oblivious to what was unfolding; asked when they first noticed smoke or flames – or Brendan – they came up with very different answers. Brendan was at the local service station at 1.16 pm. Could it be proved that he'd had time to drive to the plantation to light a fire just after 1.30 pm? His car engine was playing up and he had to drive slowly. How fast could he have actually travelled? The Arson Squad had sourced a similar car to test the potential speed, but how could they replicate Brendan's specific mechanical problem?

The second issue was McCrickard's concern that Brendan was going to be railroaded because of his psychological and intellectual impairments. A lot of people feel intimidated, even threatened, by 'strange' people. It was almost natural for the community to make a series of assumptions ending in, 'Oh well, if he was up there, it would have been him.'

The barrister thought that in preparing the brief the police had 'over-egged the omelette'. The focus should have been on what the witnesses had observed, not hearsay about hearsay. Much of her initial focus was therefore on trying to nail down exactly who had seen what, and to cut through the bias in statements made by people who thought her client was weird.

A charismatic 23-year-old off-road motorcycle racing champion had been arrested for lighting the Delburn blaze. The police alleged he set a series of fires after discovering his girl-friend had been unfaithful to him. Did he look like a textbook arsonist? As Brendan Sokaluk did? Not at all.

In May, the police recorded Brendan sounding off about their lack of progress catching the real arsonist. Kaz asked about what was by then a local rumour, that there had been someone else in the hills that day, who may be the culprit:

Brendan: 'They know who it is but they're too lazy to get off their arse and go down there and shut the whole bloody fire department down and question them properly . . .'

Kazimir: 'Did you see a motorbike up there?'

Brendan: 'Saturday?'

Kazimir: 'Yeah.'

Brendan: 'Um, I only remember seeing a vehicle that cut me and Brocky off. We were looking in the bushes for rubbish and then I saw it again . . . I drew a picture and gave it to the lawyer.'

Except he hadn't given it to her.

To be sure, Brendan's story still slid around. And as McCrickard worked on his defence, she wondered if he'd dropped a cigarette, as he claimed, or if there had indeed been someone else near the plantation that day. What if there was truth in the idiotic-sounding story he'd put in his Crime Stoppers report? She was worried that Brendan had happened to be up there near the fire and stayed around because he was fascinated by it all. During his late-night police interview, he'd mentioned having long ago been a member of the Churchill CFA – as with most of his endeavours, his parents told her, that hadn't worked out. Perhaps he'd stuck around Glendonald Road just wanting to be helpful, but instead got blamed by people who put 'two and two together'. Had Brendan believed their addition and convinced himself he was responsible?

It is criminology 101 that a defence lawyer doesn't discuss the fine details of a case with the client until the finalised brief of evidence is served by police – 'They say this, what do you say?' There's no point starting to take instructions with evolving information; the already overwhelming process only becomes even harder. And in Brendan's case it was uncertain if he was capable of ever giving instructions. If he couldn't outline his version of events in regard to the allegations against him, he would enter a whole other legal realm. Defence lawyers want their clients to be found fit to plead: the alternative to a trial is indefinite incarceration in a psychiatric facility, a fate uniformly regarded as one to be avoided.

The process of having Brendan psychologically assessed was taking its time. Not least because when the first specialist visited him in prison, Brendan, heeding his barrister's warning, had refused to talk. The next step, legally, was the committal hearing, to test whether there was sufficient evidence to proceed to trial. There was no requirement for Brendan to put up a defence at this hearing and so, while the question of his capacity was being determined, Selena kept adding information to her butcher-papered walls.

Inevitably, many of the witness statements she read reflected the community's grief and resentment. A common theme was that Brendan was a sly, calculating man. She had seen this in other cases: accusers insisting that the accused was feigning his symptoms of intellectual disability, 'putting it on'. The great social contradiction, as Selena saw it, was the general forbearance shown people with mental disabilities, and acceptance of their need for understanding – but *not if they do that*. When a mentally ill person did something heinous, goodwill vanished in an instant, and the radio shock jocks were soon telling an audience eager to believe in their own righteousness and sound mental health that the impairment was fake.

Brendan's former colleagues at the university believed he had played up his disability when it suited him. In the brief of evidence, Peter Townsend was quoted: 'One day when I was working with Brendan I told him during a conversation that he didn't seem that stupid. In my opinion, Brendan had the mentality of a student in year 7 or 8 . . . Brendan told me that when he had to be assessed he'd froth at the mouth and talk silly so the people would think he's mentally impaired. Brendan never used to try this at work when he got in trouble because we all knew he wasn't that stupid.'

Brendan's supervisor had told the Arson Squad detectives: 'Brendan is the most cunning person I have ever met. I am aware that he may have a mental impairment of sorts though he uses this to his own advantage. I felt he exaggerates the impairment. Brendan is a smarter person than what people give him credit for. He would mumble and make himself difficult to understand in certain circumstances and then speak in a clear and coherent manner at other times.'

The supervisor cited as evidence of Sokaluk's intelligence his almost photographic memory of where pipes and cables were located on campus. When underground infrastructure needed updating, the gardeners could rely on Brendan to recall exactly where equipment had previously been buried. He was also, they claimed, very technologically literate. 'He knew a mile more than anyone else about computers and phones and things.'

Again, McCrickard knew that people could find it hard to understand how someone can be adequate, or even talented, at one skill, but otherwise operate on a completely different plane. Typically, the proficiencies of someone like Brendan were the result of protracted efforts to find ways of earning respect and fitting in. They weren't markers of his average ability.

'I didn't light any fires. But I'll put my hand up, like at Monash,' Brendan had said when he was arrested, intimating he'd been wrongly accused before.

By the end of his time at the university, Brendan had begun taking antidepressants to deal with his anxiety and low moods, which were, he said, due to being his colleagues' 'punching bag'. They found him unpredictable and lazy. Once when a colleague went looking for him, and found Brendan had wandered away, he received the text message: *I have to see my doctor for my happy pills so I can be normal like use.*

The gardeners believed that because Brendan had been employed as part of a special program, he could not easily be fired. If they complained about his behaviour, they claimed, his mother would turn up and accuse them of bullying. (Selena thought of the exhaustion Lou Sokaluk must have felt, marching in yet again to protect her overgrown child.) Further up the ranks, the university's administrators were not greatly interested in the garden staff's squabbles.

The supervisor told the police he had to be careful about where on the campus he sent Brendan to work. He'd been seen peering in the windows of student dormitories, so during term time he was assigned to other places, although not the child-care centre, because the mothers apparently found him creepy. He couldn't work on the golf course, because golfers complained that he picked up balls in play, or ploughed over them with the mower. As a result, he was banned from driving on the campus.

The supervisor had prepared a list of Brendan's disagreeable behaviour to give to Monash's human resources department. It included:

- Deliberately writing with his non-preferred hand so that his script is illegible.

- Following colleagues home and informing them later 'watch out, I know where you live.'
- Passing on a note to a female colleague challenging her boyfriend to a duel.
- Trying to claim compensation suggesting hearing loss until he was found out.
- Openly shows happiness when a colleague might be grieving at the loss of a loved one.
- Hiding things in difficult to reach places causing annoyance to colleagues. For example, a colleague has stripped a machine for repairs and is now ready to rebuild, but a vital machinery component is removed by Brendan.
- Disturbs loose ground directly upwind of other workers with the result that they are showered with dust.
- Mows over rubbish (thus making a mess) rather than picking it up.
- Is observed at work functions standing around the tables stuffing food into his pockets.
- Consistently climbs into the industrial waste bins to look for useful rubbish to take home, during work time and in contravention to his supervisor's instruction.
- Constantly tells stories that are totally incomprehensible.
- Constantly makes cat noises 'meow, meow' directly at people.
- Loves to work in the mud when its raining, diverting surface water into open drains.
- Lifting the guard of his ride-on mower, thereby showering debris on to his colleagues.

- Deliberately running over golf balls and causing the golf balls to propel towards any bystanders creating potential safely risks.

It was hard, after eighteen years, to work out who had launched this epic quarrel or if there was any way to end it. But Brendan reckoned he'd been blamed for things he hadn't done and his fellow gardeners reckoned he wrought mayhem. They'd grown weary of babysitting someone so childish: not a funny, likeable naif – the daytime-movie version of an intellectually disabled person – but this ill-kempt, complicated man who left a trail of trouble.

On 7 June 2006, Sokaluk and his supervisor – the man who'd compiled the list – had their last run-in.

Later, the supervisor claimed he had assigned Sokaluk certain tasks and returned to find that 'he had just downed tools and walked off'. He then received a text message: *I,m going to comit suitside by taken a lot of pill,s so I can die and not put up with youes.*

The supervisor contacted the university's security team, the police, community services, Brendan's parents, and the local hospital. Eventually, Brendan turned up at his doctor's office. He told the doctor he had been driving along winding Jeeralang West Road looking, without success, for a logging truck to crash his car into.

Brendan's family urged him to put together an account of his last day of work, which he emailed to the university:

Subject : this exsplains what happen and why I losted.

[My supervisor] had a angry look on his face I was worried
for my safety again. He told me to get in the ute. I got into
the ute slightly sceared of what he was going to do towards
me again. He drove donw to the sports field on the way down
he was yellen at me. Sayen this. When I fucken tell you to do
something I actspect you to do it and not tell me your fucken
girlyback herts when sitten on the gator [tractor] seat what
are you a fucken girl I,m sicken tired of putten up with you if
I had m y way you will not be here at all. As I look across to
him he seid to me I,m going to smash your fucken head in do
you hear me you fucken asswhole. I kept quit then he seid to
me. You fucken retart answer me now. I kept quit. He drove
the ute very farst over the kerb and stop at the sport field . . .
He also stated no one likes you and no one whants you here
because your a dumb fucken retart . . . you fucken do what I
whant or you whount work here any more. I steard at him. He
turn and walked of I watch as he left I stood there for awile
then I sat in the sun on acorrect ledge. I had no one to talk to
about this I felt alone and sad. I got up and chip some weeds
and prun a bit for awile. I got sader and sader then I decited
to leave. I sat in the car for a while. I was so sad I started to cry
about what happen . . . I decide to flee for the hills to find a
mack truck to smash my car into. Un forgerly I could not find
one. I drove up and down many dirt tracks looken for one.
Arther a while I setile down and deiced to go to the doctors to
get help. Now I,m of work on stress leave and safe away from

[my supervisor] he can,t get me here. I SWEAR THIS TO
BE ALL TRUE OF ALL THE AVENT THAT TOOK
PLACE. FROM BRENDAN SOKALUK. I all so have
noterfived the uoin over this matter.

His supervisor disputed this record of events. However, after years of mutual grievances, the university wanted Brendan Sokaluk's employment terminated and he was paid out.

His world now retracted further. He might have been impaired, but not to the extent that he couldn't recognise other people's contempt for a 'fucken retart'. He searched for jobs on the internet, which also offered the easiest place for him to socialise. Online, he met Alexandra, via her Myspace page.

Alexandra worked in a nearby aged-care facility, and in her profile picture – the one later plastered across the internet, showing her in a CFA uniform and angel wings – she stood in a verdant Gippsland garden, a tiny view of a power station's chimneys visible behind her. As an infant, she claimed, she'd fallen into a bath and suffered severe burns to her face, arms and torso. When she grew older: 'kids was so mean to me at school. I was outsider. I could never fit in . . . it hurts rest of my life what they did.'

The pair chatted online for two months before they met. Alexandra was also a vulnerable person, who lived with her parents and needed their help navigating the world. Brendan would pick her up from their house and drive her back to Sheoke Grove. The new couple watched movies: 'action and war and violence DVDs . . . Brendan did not like the sort of comedy and girly DVDs that I liked,' she later told police. Nevertheless, within weeks she had moved in with him. 'I thought we were in love.' One day she announced on Myspace: '[t]he princess is getting married,' and Brendan responded, 'I love Alexandra more than anything else.'

In their brief of evidence, the police listed the various web searches they'd found on Sokaluk's computer: 'BRAIDLE DESS COURSES' – Alexandra was dyslexic – and links to advice on wedding planning, hen's night celebrations ('my spa party – indulge yourself with home facials'), eloping, and luxury honeymoon accommodation.

Brendan's father still came over to clean the house three times a week. As Lou was working, he'd acquired the day-to-day tasks of watching out for their son. He'd take Brendan grocery shopping, deal with his bills and paperwork, and dole out his allowance. (Brendan told Alexandra that someone he'd trusted at the university had touched him inappropriately and he'd been given a payout.) Meanwhile, the couple started watching their weight and attending literacy classes at the local community centre. They bought a dog from the RSPCA: a small, tan-coloured terrier. Despite the dog being female, Brendan named it in tribute to the recently deceased racing car driver, Peter Brock.

Brendan and Alexandra shared a bed, but in the three or four months they lived together they never had sex. Nevertheless when Brendan was told by another internet acquaintance that Alexandra had recently slept with someone else, he was enraged. According to Alexandra, 'He kept saying, "How could you?" and calling me a whore.' About a week later, she called her parents, who came and picked her up. Brendan changed the locks.

That had happened a year ago.

Yes, he'd been angry at the time, but it hadn't seemed to Kaz and Lou that he'd spent the next twelve months desperately lovelorn, or hell-bent on revenge, as the newspapers made out. He had his own quiet routine. And routine was very important to Brendan. Although it might have seemed bizarre to outsiders, it hadn't struck his parents as strange that he was up at the plantation during a heatwave. His days revolved around a limited set of activities, such as searching roadsides and gullies for old scrap metal, tinkering in his shed and gardening. He'd grow vegetables in the planting beds that were free of junk, and give his produce to those he considered friends. And the junk often had some value. The Sokaluks believed he burnt things in his yard not, as the media intimated, because he was some sick pyromaniac, but to salvage the precious metals: copper inside electrical cords, and gold from computer circuit boards. It was economic, not pathological.

Twice a week he did his paper round. He went to bingo, and looked forward to car expos, or visiting the Rosedale Speedway with his father, where they watched Crash n Bash V8 super sedans racing. He tried to keep up with the literacy classes even after Alexandra left, and he took the teacher vegetables to thank her for her help. This teacher never saw Brendan angry or confrontational. If others in the class argued he'd sit back and wouldn't get involved. She was another who lost her house in the fire.

Neither Kaz nor Lou could believe he'd lit the blaze. Each day there was another 'revelation' about what their son had supposedly done, and they tried to navigate between what seemed truth and the sharp turns required to look at the rest.

Selena had sat in interview rooms with a lot of families whose children were accused of despicable things. Many parents were just doing the best they could, and that wasn't always a great deal. Often they were angry or depressed, having been ravaged by misfortune themselves, and couldn't stop thinking that their own lives were hopeless. She thought the way Kaz and Lou managed to care for Brendan was an achievement in itself. In this regard, at least, Brendan Sokaluk had been lucky.

When at last, in July, six months after the fire, the police handed down their final brief, it totalled more than 5800 pages, with hundreds of supplementary documents. Brendan faced ten counts of arson causing death, and 181 other charges, the majority relating to criminal damage. He was not charged with the death of 86-year-old Gertrude Martin, who suffered a heart attack while her house burned down. As she'd had a pre-existing heart condition, it was too difficult to prove that the fire was the sole reason she died.

Soon Selena McCrickard was moved to another department within Legal Aid and reassigned to other cases. She'd given so much energy to Brendan's defence that she was gutted by the change. Later, Lou Sokaluk asked for her to be reinstated, but Legal Aid's managers already had a legal team in place.

In the period it took to determine whether his case would reach trial, Brendan cycled through a number of lawyers. It had become clear to the Sokaluks that the law could warp the physics of time. Their son's arrest had happened at full speed and yet the legal process seemed never to get out of a crawl. They were battling now over the future, the present, the past, in torturous slow motion. Meanwhile, they worried about Brendan coping in the maximum-security jail, who he'd be meeting there, whether he'd be safe.

Selena did not often stay in touch with her clients, but she'd become close to Lou. She wanted her to know that someone understood how crushing the legal process could be for *all* the families involved, and that she cared beyond the hours that were billable.

She visited Lou when, fifteen months after the fire, in June 2010, Brendan's committal hearing began. The older woman's hair, she noticed, had turned from dark brown to white. During this proceeding, Lou heard about the family that lost their pet horse. Lou had been a passionate rider as a girl, and that loss symbolised for her every other. She told the barrister she wished she could afford to buy those people a new horse, but she didn't have the money.

Defence lawyers, Selena often wanted to explain, don't set out to say, 'Look at my client, he's a fucking great bloke.' Rather the point was: 'Look at my client, who may or may not have done this terrible thing. This is the issue with this person, and he could be your child.'

The Sokaluks had now spent the greater part of their lives as 'the parents of the retard'. Kaz had ignored the looks people gave them when Brendan babbled on in public in a foghorn voice; Lou had fought over and over to protect her son from all those she felt deliberately gave him a hard time. She wasn't going to stop now. It was devastating to think he could have set the fire, and to know that most of their neighbours believed he had, but he'd been picked on all his life: what if this was no different? Until she heard Brendan say he'd deliberately lit up the plantation, it was a betrayal to believe it. Until her son told her he'd done it, she wouldn't stop trying to defend him, even if she lay awake wondering.

In the meantime, Lou said to Selena, what about poor Ron Philpott?

On Black Saturday, Philpott, a local CFA member and 'fence hand', had been the first person to report a fire near an old mill close to his run-down shack in Murrindindi, a rural region a hundred kilometres north-east of Melbourne. Before long, the Arson Squad considered him a suspect. This shabbily dressed man with his sunken face and busted up nose was soon on prime time television, proclaiming his innocence.

Forty people had died in that fire, and the suspicion cast upon him made his life hellish. But after a thirty-month long investigation, the Arson Squad quietly announced, in June 2011, he was no longer a person of interest. They'd concluded that the most likely cause of this fire was a faulty powerline, as Philpott had told the police from the start.

What if Brendan had also been in the wrong place at the wrong time, and falsely accused?

Eventually, nearly three years after Black Saturday, following a lengthy, stop-start process of psychological evaluation, Brendan was deemed by various specialists competent to stand trial. He was charged with lighting the fire at age thirty-nine. He was forty-two before his court case began.

During the trial, Selena was heavily pregnant with her son.

When you hold your first baby and you imagine the life they might lead, you don't foresee the infant growing but not growing up. You don't picture a child-man, approaching middle age, still sending out birthday invitations to the few people who tolerate him, with further directions scrawled on the bottom: 'Bring lots of lollies!' 'Bring me a present!'

You don't foresee yourself visiting him in jail, and his telling you that he'd prefer to talk through glass because after 'contact visits' the guards make him strip down and spread his buttocks to check for contraband.

You don't predict the ferocity with which people will come to despise him.

Or that you'll suffer nightmares about an inferno he is charged with lighting.

Selena believed that a lesser person than Lou would have crumbled under the same pressure. The barrister hoped that if – fate forbid – she was ever pushed to the place this mother had been, she could summon some of her grace.

On Valentine's Day 2012 a group of defence lawyers stood watching a 'jury view'. The jurors were inspecting the edge of the plantation where it was alleged Brendan Sokaluk had deliberately lit two fires. In the summer heat, on the fouth day of his trial, they gravely made notes, and gravely considered the possibility that they were standing at the scene of an appalling crime. A haze of insects drifted in the gaps between neat rows of trees.

Ten thousand hectares of plantation timber had burnt in the blaze: the *Eucalyptus globulus* destined for the local paper mill, the long-fibered *Pinus radiata* for the timber mill. The debris had all been bulldozed and the land replanted. Here, by the dirt track named Jellef's Outlet, the gums had been replaced with pines. The young ones were like Christmas trees, but on the other side of Glendonald Road, the stand of unscathed *radiata*, 26 years-old and storeys-high, remained as silent witnesses to the fire's beginning, three years and one week earlier.

Jarrod Williams, a Legal Aid public defender acting as Sokaluk's junior barrister, stood back assessing the jurors. It was a professional pastime, trying to size them up without their noticing. As the horrors of Black Saturday had entered the public consciousness, he assumed they had surely entered the consciousness of these people too. The country's worst recorded natural disaster had changed the way even city folk thought about summer. These strangers, their lives suddenly disrupted, wore fluorescent jackets, as did the judge, who looked pleased to be out of the Melbourne courtroom. A minibus parked to the side of the road gave the jury view the feel of a school excursion – except these men and women were responsible for deciding the fate of the one man whose Black Saturday arson charges had reached trial.

Williams could see the geography of the story dawning on them. The fire's path was still evident on the hills adjacent to the plantation. There were clusters of bone-white eucalypts where the inferno had burned too hot and too deep into the ground for the trees' survival. The steep angles of the land meant the damage was on permanent display. These sloping hills, the heat, the quiet, the ease of ignition, and of finding oneself trapped – none of it was hard to imagine.

Courtrooms disembody the evidence. Melbourne's Supreme

Court building was a grand, goldrush-era edifice; all the embellishments of the high-Victorian decor matched the intricate legal formalities. But now the main players of this drama, other than Brendan himself, stood in the dirt, with birds darting and calling in the taller branches overhead.

The jury were shown one proposed ignition site by detectives Adam Shoesmith and Paul Bertoncello, then they navigated the new pines to the second site, a hundred metres away, on the other side of Jellef's Outlet. They were followed by the Honourable Paul Coghlan, the presiding judge, without his wig and gown.

Ray Elston, a veteran prosecutor with a calm, even manner, stood with his junior barrister and solicitors. Jarrod Williams stood in a separate group with Brendan's instructing junior solicitor, Gaby Pulczynski, from Legal Aid; and Jane Dixon, the star silk, also provided by the state, who was leading the defence.

From their point of view, this trial was going to be difficult but was far from unwinnable. Their case, outlined by Dixon the previous day, was this: after an event 'of catastrophic significance', Brendan Sokaluk was 'an easy target. So easy that he even managed to make himself believe that something he did must have caused the fire in Glendonald Road.'

Now they were standing by Glendonald Road, with Shoesmith and Bertoncello, whose investigation Dixon had excoriated in her opening address. After working on this case off and on for three years, these two officers had become personally invested in Brendan's conviction. Things were rarely relaxed between a defence team and the police, but the atmosphere here had an undertow of aggression. In short, the defence lawyers sensed that the police detested them.

There was a stand-off, already established in the opening days of the trial, over who Brendan Sokaluk actually was. The police and prosecutor's view that he was cunning and calculating vied with Dixon's story of a hapless naif, a simpleton more sinned against than sinning, caught up in events beyond his control.

Jane Dixon, in her late forties, had an enigmatic part-smile and a sharp-faceted intelligence. She was at that time only a few years away from becoming a Supreme Court judge herself, and was the president of Liberty Victoria, a platform from which she'd championed human rights, particularly of the disabled. As a barrister she had a gift for understanding who her client was and building a narrative like a raft around them.

But a corollary of Sokaluk's innocence, in her view, was the Arson Squad's incompetence, and Dixon had criticised the forensic investigation of the ignition sites, suggesting there had been 'helicopter bombing, bulldozing, raking and hosing' in the area, which had disturbed the alleged crime scenes.

The defence had also recruited a well-known wildfire expert, the academic Kevin Tolhurst, who in the months after Black Saturday had achieved almost celebrity status due to his regular television appearances. He prepared a report damning the investigators, including for not giving due consideration to the possibility that the second fire was actually the result of a spot fire. Tolhurst believed that localised wind effects, turbulent eddying and spiralling gusts seen only in extreme fire conditions, could well have ignited the second area of the blaze's origin.

More controversially, Tolhurst also wondered if another arsonist had been roaming the hills. In her opening address, Dixon had flagged his claim of a suspicious fire that he called Churchill Two, just over a nearby ridge. Churchill Two, his theory went, had been lit after the two other fires were already raging on Glendonald Road, and was missed by the Keystone Kops.

Adding insult to injury, Dixon now asked Detective Shoesmith to show the jurors the site where Tolhurst suggested this Churchill Two fire had been ignored.

What if, the defence intended to suggest, all the while that Brendan Sokaluk had been trying to start his car, believing he was somehow responsible for the burgeoning inferno, the real perpetrator had got away? The rumours of suspicious motorcyclists in the hills on Black Saturday were ongoing, to which Tolhurst's evidence would add force. The defence planned to call as a witness one motorcyclist whose registration was taken down by police shortly after the main blaze took off.

'If there was an arsonist involved in and around Glendonald Road, was it the accused, or was it perhaps some other more mobile person?' Dixon had earlier asked the court. '[T]he question for you,' she told the jury, 'is, "Was this a case of arson at all?"'

The thinking around who bore responsibility for Black Saturday had significantly shifted. The Bushfires Royal Commission, an exhaustive, fifteen-month inquiry into the fires, had revealed that deep levels of bureaucratic and corporate incompetence played a major role in the tragedy. All over the state on 7 February 2009, despite fire scientists, including Tolhurst, accurately modelling where the multitude of blazes would spread, barely any information filtered through to the public. The spurious orthodoxy adopted by the CFA's senior management was that more people might die fleeing fires than if they sheltered in the houses they were trying to save. The royal commission proved the theory baseless.

The commission also decided that privatised and often outdated power infrastructure was largely responsible for damage estimated at 4 billion dollars. The Kilmore East fire – in which 119 people died, 1242 houses burned down, and 125 383 hectares were blackened – was the result of a 43-year-old conductor between two rural powerlines failing, and discharging plasma at a temperature of 5000 degrees Celsius near vegetation. (The electricity in those malfunctioning wires most likely came from the Latrobe Valley.) Now a $494-million class action was winding its way through the courts, brought by the fire's survivors. And as it became clear that a faulty powerline, rather than Ron Philpott, had caused the Murrindindi blaze, a $300-million class action was launched to follow.

It was reasonable for Brendan Sokaluk's lawyers to ask whether the public fear of arsonists contained an element of hysteria, and if the police had echoed and amplified this fear and loathing by bearing down on Brendan. Here was a man, Dixon claimed, easily scapegoated under the 'pressure to somehow make someone responsible for a catastrophic occurrence'. Their client was a Philpott-like figure, who 'obviously slipped through the cracks in the system, in Churchill', and was considered 'a bit retarded', or 'an odd sort of fellow', 'a man who in the olden days would have been described as a bit of a simpleton'. He liked 'tinkering in the shed like many Aussie blokes, except that in this case there is evidence [he] was overheard by neighbours . . . listening to *Bob the Builder* and *Thomas the Tank* tapes'.

There was also another layer to Brendan's story.

In her opening address Dixon had warned, 'It's not good enough for you, the jury, to simply say, "Oh, well, Brendan Sokaluk himself seems to think he must have caused the fire, so that's good enough for us." Why is it not good enough? In this case, you're dealing with a person who suffers from autism.'

When Jane Dixon took over this brief, there had been something about her new client's walk that made her wonder if his proprioceptive senses, his feel for his own body in space, weren't skewed. She also noticed that Brendan never seemed to recognise her. Whenever the gap between meetings was more than a few weeks, he greeted the legal team as if for the first time. Facial blindness, or prosopagnosia, is not uncommon amongst those with autism spectrum disorder; light, sounds, smells can often take precedence over facial features or details such as eye or hair colour, height and weight.

Dixon's instinct was proven correct. Brendan was diagnosed by three doctors as being on the high-functioning end of the spectrum, but also on the borderline of intellectual disability, with his poor verbal comprehension placing him truly in the range of the disabled. (He was therefore going to look more impaired, one psychologist later explained to the court, than someone with more extreme autistic behaviour who might have greater intelligence, and be therefore able to learn to compensate.) His condition had affected nearly every aspect of his schooling, employment and relationships. Why not also his relationship to the fire?

A few weeks before the trial began, Brendan's legal team had taken him back to Churchill, and the same corner of the plantation where the jury would stand, picturing the fire's start. Two months earlier, after 1041 days in custody, Brendan had finally been granted bail. One of the conditions was that he not return to the Latrobe Valley without his lawyers.

It was a long drive out of the sprawling suburbs of Melbourne. Jane Dixon was at the wheel, Jarrod Williams in the passenger seat, Gaby Pulczynski and Brendan were sitting in the back.

With his hair grown out of its buzzcut, it was more obvious that Brendan was greying. The lines around his eyes had deepened, while his face was fuller. In jail he had started lifting weights, but on bail he could easily access his favourite junk food and had added bulk.

Living with his days free had been an adjustment. For the past few years, Brendan had resided mostly in a maximum-security unit for intellectually disabled prisoners where every aspect of his day was scheduled.

The unit ran a horticultural program, and in the car Jane Dixon did her best to draw Brendan out, asking him about the greenhouse where plants were grown from seed, the ornamental garden, the orchard, the vegetables he'd planted. Skilled as Dixon was, she could not engage him in conversation beyond a narrow range. In volume five of the *DSM* the diagnostic criteria for autism spectrum disorder, which the defence generally recognised in Brendan, included, 'failure of normal back and forth conversation', 'abnormalities in eye contact and body language', and 'deficits in understanding [others'] use of gestures'.

But these are outward signs of a more profound difference. Many neurodiversity advocates prefer to frame autism as a radically distinct way of being in the world. The celebrated writer and activist Temple Grandin, herself autistic, writes of 'specialized brains' of contrasting, not lesser, intelligence.

Grandin also draws attention to the sensory minutiae of everyday life that can pose a threat to those on the spectrum, regardless of their intellectual abilities. 'A person with autism has hypervigilant senses,' she explains; 'my nervous system was constantly ready to flee predators. Insignificant little stresses caused the same reaction as being attacked by a lion.'

Many other autistic people also report living in a distorted sensory realm. 'It often felt like the effect one gets in a 3D movie where you duck and weave as everything seems to be coming at you, invad[ing] your world,' the writer Donna Williams observed. And according to Naoki Higashida in *The Reason I Jump*, 'It feels as if the ground is shaking and the landscape around us comes to get us, and it's absolutely terrifying.'

Commonplace sensations can be transfixing and distressing: Grandin describes certain clothes feeling 'like sandpaper scraping away at raw nerve endings', rain sounding 'like gunfire', a fluorescent light 'flicker[ing] like a disco', a balloon popping like an explosion, crunchy food that's too loud to eat, a 'hair dryer like a jet taking off'.

Many autistic children deal with delayed speech. In *Thinking in Pictures*, Grandin quotes Donna Williams' claim that as a child she 'heard speech as only patterns of sound'; Jim Sinclair recalling he had 'no idea that this could be a way to exchange meaning with other minds'; and Darren White's difficulties communicating because 'another trick my ears played was to change the volume of sounds around me. Sometimes when other kids spoke to me I would scarcely hear, then sometimes they sounded like bullets.'

A lack of eye contact can be due to what Grandin refers to as 'sensory jumbling', an inability to look at someone and listen to them at the same time. For Donna Williams, it was hard to concentrate on speech while focusing on a face, 'because the eyes did not stay still'.

In other words, Brendan Sokaluk likely inhabited an alternate reality. He spent much of the car trip asleep, but woke when they were close to the landscape of his life before the fire. His 'specialised brain' included a savant-like visual memory. He was alert to the topography of the Valley in ways that would impress the most experienced surveyor. And yet, back on the map he knew best, as the car passed the giant open-cut mine and, after the turn-off to Churchill, the industrial zone that led to the quiet town, Brendan stared impassively ahead and it was impossible to guess what he was thinking.

Next to him, Gaby Pulczynski was growing used to Brendan's seeming blankness, and the complexities it may have hidden. The legal team had now spent a great deal of time with him. They'd needed extended sessions because of his scant attention span and verbal comprehension. Often the lawyers had to give the same simple information over and over, and ask him to repeat it, to check if, in some form, he had followed. Whether or not he had was frequently unclear.

With his family two and a half hours away, Gaby had to give Brendan a lot of the help he needed. Kind by nature, she'd grown almost fond of him. He would try to make jokes, and even though they weren't funny she'd laugh for the poignancy in the attempt. She'd buy him a packet of chips from the prison vending machine because he'd be as excited as a kid at Christmas. And after he was granted bail, she found herself, along with Shannon Dellamarta, his senior Legal Aid solicitor, escorting him to a billeted assisted-living facility and making a detour to his beloved KFC. Checking the weather each morning, she reminded him on days of total fire ban, anywhere in the state, that he was forbidden to go out. Brendan, meanwhile, remained adamant he had never deliberately lit a fire. The young lawyer wanted him to have the best chance to prove his innocence.

Pulczynski was impressed by Jane Dixon's lack of ego and her dedication. When they arrived in Churchill, Dixon, a non-smoker, stopped at the petrol station to buy a packet of cigarettes. She lit one up as she drove from the town: a form of legal method acting. Past the golden cigar, gleaming in the summer heat, they approached the plantation with the coiling smoke drifting out the window into the breeze. She wanted to see how long it took for a cigarette to burn down and how far she could drive while it did. But it was less a scientific test than a means of jogging Brendan's memory.

And then there they were. Near the intersection of Jellef's Outlet and Glendonald Road. The trees quivering with insects, sap blistering the bark in the warmth of the day. And there he was, Brendan Sokaluk, telling them again in that slow, slurring voice about catching cigarette ash in a napkin. His face numb while his hands, as he spoke, were in repetitive movement, finger and thumb rubbing together or lightly touching the fabric of his clothes.

What Brendan was thinking at being back here, beyond the mechanics of his story, was again impossible to tell. Perhaps he was remembering his many happy days scavenging in the forest. Or the sky-blue car in which he cruised around the service tracks. Or the drama of that day when he walked through the fire with Brocky.

He'd lived near these hills for most of his life, though he'd never known the virgin forest in which they'd once been cloaked, or heard the lyrebirds that added the sounds of axes and chainsaws to the repertoire of songs they passed from one generation to the next. He'd told the detectives the forest was a special place for him – '[the] one thing I loved in my whole life'. But there was nothing in his reaction now to suggest that he was moved at being returned to the remnants of it.

Temple Grandin, upon seeing the joy the psychiatrist Oliver Sacks found in a sunset, had confessed her sadness: 'I wish I did too. I know it's beautiful, but I don't "get it".' She believed that while stargazing she 'should get a "numinous" feeling', an almost spiritual sense of awe, but it didn't come. And despite knowing all the names of the birds and plants and geological features of the countryside near where she lived, Grandin claimed to have 'no special feeling for them'.

Brendan, like Grandin, seemed to experience his environment in utterly different ways. He could hold the nuance of the area's complicated geography in his head – the lawyers soon realised he knew all the unmarked tracks through the plantation – but back in this place where his life had gone awry, he was nonchalant.

All of us can turn away from the consequences of our actions. The effects of the cigarette that may or may not have fallen, the lighter Brendan did or didn't put to the leaf litter might have become too overwhelming for him to consider. Or maybe his brain did not make this connection.

In the brief of evidence, the transcripts of telephone calls between Brendan and his father included one recorded a few weeks after the fire:

Kazimir: 'So, you still reckon you didn't do it?'

Brendan: 'Yeah. I know I didn't do it. I think it's the same as like Monash. They – you, you tell 'em what, what really happens and they just change it, you know?'

Kazimir: 'Yeah.'

Brendan: 'After a while and you remember, "hang on, that's not how it went."'

Kazimir: 'No matter what, Brendan, you can't come back here.'

Brendan: 'No. I'm not coming back.'

Kazimir: 'You have to go somewhere else, some other town . . . change your name and shit . . .'

But later, Pulczynski walked down the main street of Morwell with her client and stopped to buy lunch at Subway. If anyone recognised him, no one said anything. The man who three years earlier the locals had wanted to knife and shoot and burn was hidden in plain sight. Brendan stood staring at the pictures of bread rolls stuffed with cold cuts and seemed to fit right in.

Except that he didn't. The lawyer noticed that even the task of ordering a sandwich was complicated for him. Those with autism spectrum disorder are sometimes said to have no 'theory of mind'; that is, a deficit in the ability to recognise that other minds think differently. It wasn't just that Brendan had difficulty reading the menu, he was also unable to read basic social cues; he didn't register the shop assistant's mouth set in waiting position, or the inflections in the usual back-and-forth of ordering. 'Yes?' the assistant kept asking, becoming irritated. 'What do you want?'

That there was something disconnected about Brendan became more apparent to Pulczynski when they were out in the wider world. It was as if the glass partition of the prison interview room went with him: there was always some invisible wall that separated this man from those around him. He lived in an echo chamber. He couldn't really 'hear' what was being said to him, and it was difficult to make out what he was trying to say.

All this blurred him. 'His features are nondescript,' one witness had told the police. 'He has just the face you sort of recognise in the car. It's not the face I say *G'day* to if I walked past, or anything like that. Just a face you wave to in the car as it passes.'

Brendan's lawyers knew there was a chance the court would hear a simplistic version of his history. Allegations in the brief of evidence suggested he may have lit fires in the past. But there could also have been a folkloric element to these stories: the hearsay of hearsay that in small towns rises up from the substrata. In the days when Brendan was so despised, it was no doubt easy to unload gossip about him down at the police station.

Someone had once seen a strange expression on his face while he used a blowtorch in a high school woodworking class. Someone else remembered a group of teenagers, including Brendan, setting a fire in the toilet of a local milk bar and, later, in a park at the back of the high school. 'The fire cleaned out the whole park,' recalled this former acquaintance – another school outcast. 'About 10 acres went up in smoke. I do remember Brendan saying that he liked that the forest was now clean of all the rubbish, like how the Aboriginals used to do it.' This man's story kept changing, however, and the defence found out that he had been to court in the past for dishonesty offences.

After he left school, Brendan volunteered with the Churchill CFA. In June 1987, he was accepted for active membership. Some time the following summer, on 'a really hot, terrible day', a family friend of the Sokaluks noticed a fire in a paddock she was leasing for her daughter's horses. The two of them rushed towards the flames.

When we got to the paddock I saw the fire brigade coming. When the fire truck pulled up [Brendan] was on it. He was dressed in CFA gear with his CFA jacket. The fire was just in the bale [of hay] and the CFA put it out. Once the fire was out we saw a box of matches on the bale. The fireman took the matches straight away. I thought it was Brendan who lit the fire because he'd just joined the CFA, loved the trucks and the power of being a hero. He would have only been 19 or 20 then but had the mentality of about six. I can't say it was him but I just know it was . . . I didn't call the fire brigade at all so I don't know how they found out.

In early 1988, Brendan was dismissed from the CFA for 'dishonest conduct'. The Arson Squad had located the CFA captain of the time, an ex-policeman now working as a bus driver in Queensland. He claimed that Brendan began appearing at a series of fires to which he hadn't been called. After a lengthy conversation, Brendan allegedly admitted to lighting the fires so he could go out on the truck and help fight them.

But Lou Sokaluk told the lawyers that all those years ago, the CFA captain, an officer of whom the family weren't fond, had not told Kaz or Lou about any lengthy conversation with their son, nor any admission on his part. They'd been led to believe that Brendan was dismissed from the CFA for a different transgression.

One evening around that time, Brendan had entered the Churchill high school after hours, setting off an alarm. Lou

Sokaluk acknowledged that he shouldn't have been there; it was stupid, the wrong thing to do. But her son had trouble with basic reasoning, with common sense. He did dumb things. Why, she wondered, if he'd been lighting fires, had the officer never come and told her and Kaz? If half the town now claimed they suspected he was a firebug, why had no one ever mentioned it to them?

Years later, after leaving his job at Monash, Brendan had tried to join the Churchill CFA again, but his mother claimed this was only at her urging. She wanted her son to have people to talk to, a place to go, a task of some description that would fill his very lonely days.

If Brendan was asked to whom he felt closest, he would answer, 'Brocky.' Even after their long separation, he spoke of the pet often and in terms that could be confusingly human. He believed, for instance, he knew which chips Brocky preferred from the different junk food outlets they'd frequented together.

Dr Marged Goode, a clinical psychologist specialising in autism spectrum disorder, would later give evidence at the trial. Of his devotion to the dog she told the court, 'Pets are actually very good at reading your body language and it doesn't matter if you can't read theirs. Pets don't tend to be sarcastic or make jokes or do any of the things that people do that [those on the spectrum] find so hard to deal with.'

Brocky had often been the only thing stopping Brendan from being totally friendless. And so, on their visit to the Valley, Jane

Dixon arranged to meet Kaz Sokaluk somewhere discreet so Brendan could briefly spend a moment with his dog. They all assembled on Cemetery Avenue, between the power station and the Hazelwood pondage.

As Brendan and the legal team stepped from the car, there was a sense he almost enjoyed their company. This was perhaps the first moment in his life that a group of people had spent time asking him questions, listening to his stories and asides, accepting him, more or less, for who he was. He was reunited with his pet, and the lawyers stood back to give him some privacy, half distracted anyway by the view.

This was a surreal world for those not used to seeing the gargantuan infrastructure that makes our electricity. The towering chimneystacks looked like the columns of a temple, the veil of thin brown smoke a reminder that fire here was constant.

Brendan wanted to hug Brocky, while his father, perhaps trying to protect the animal from Brendan's overwhelming attentions, preferred to keep the now renamed dog at a distance on the lead. Even this interaction, once Brendan's mainstay, had become complicated.

Part III

the courtroom

Each morning of the long, tense days that followed the jury view, one of Brendan's solicitors would call him at the assisted-living facility to remind him it was time to leave for court. He had rarely been to Melbourne as an adult, but his lawyers had shown him on a map the simple tram route into the city. They would meet him at the tram stop or the Legal Aid offices, and escort him through the columns and arches of the Supreme Court.

Shannon Dellamarta had drawn a picture of the courtroom so Brendan could visualise where he and the jury and the public – some of whom had lost family members in the fires – would be sitting. She added the raised bench opposite him for the judge, who in his wig and gown Brendan believed looked like Father Christmas.

Dellamarta and Gaby Pulczynski, as the instructing solicitors, sat across from the barristers, facing the room, and keeping a weather eye on Brendan. During his committal hearing he had occasionally fallen asleep. When Brendan entered the dock, one of them would hand him a notebook and coloured pens with which he'd set to work drawing, obsessively alternating between red and blue. The police felt this was a ploy to make him appear harmless, but the defence hoped that if Brendan was occupied he'd be less likely to doze, or fool around.

The lawyers were also worried about his tendency to openly yawn. It hardly mattered that the combination of his medication and his short attention span was the cause, yawning looked like brazen indifference to the proceedings, and the disaster at the centre of them. And Brendan *was* bored. Believing his lawyers were too, he took to feigning sleep and waking as if startled. They firmly told him this hamming had to stop.

On the fifth day of the trial, Kaz Sokaluk appeared as a witness. Kaz was a reserved man who tried to live a quiet life. He had been conceived while his Polish mother was interned in a German labour camp, and he had never gone back to Europe, nor ventured, in the past forty years, far from the Valley. It was therefore a considerable effort of will to climb the small wooden steps to the grandly carved witness stand and wait there, belittled by the opulence of the high-ceilinged room, conscious of every tremble in his hands or voice.

The trial travelled back in the imperfect time machine of personal recollections. 'If I could take you to 7 February 2009, the Saturday,' prosecutor Ray Elston said. 'Did you have a regular routine with Brendan on Saturday morning?'

'Yes.'

'And this one, you went into Morwell with him in the morning?'

'Yes.'

'And ended up going to Kentucky Fried for lunch?'

'That's correct.' Kaz had seen Elston at various legal proceedings over the past few years and the prosecutor had always been respectful.

'Dropped into Autobarn?'

'Yes.'

'And Supercheap Auto, just to have a look around?'

'Yes.'

The court heard that father and son then went to the TAB to lay a few bets. Then on to Big W, and the hardware chain Bunnings to check for bargains, before stopping for lunch. During the week, Kaz explained, he would visit Brendan's house and check that it was clean, making sure also that his son had enough food and money. He told the court how Brendan supplemented his pension with the paper round and by collecting scrap metal and doing odd jobs, cutting firewood, mowing lawns. Then, on weekends, they had this regular outing.

But in the heat on Black Saturday, Brendan's car 'was losing power. It was jerking, backfiring. It just wasn't running right.' As they sputtered back to Churchill they passed a fire engine attending a grass fire near Energy Brix, which produced coal briquettes. They 'stopped to have a look, same as everybody else'. Brendan dropped his father home and said he was driving up towards the hills to see his friend and ex-neighbour Dave. He needed to pick up a chisel set he'd loaned him. Brendan's house didn't have air conditioning and he 'reckoned it was cooler up there. He's often been up there when it was hot because it's up in amongst the trees and that.' Kaz told him not to go, that he'd tune the car up the next day, after the cool change.

Lou Sokaluk wasn't in the gallery watching her husband give evidence. The two of them shared a train ticket and alternated their days in court to defray the expense and distress. It felt to Lou, the lawyers believed, as if everyone were sitting in judgement of her and Kaz too – and the whole town seemed to have been bussed in to give evidence. She'd had to accept there was much about her son she hadn't known: that he had no real friend was perhaps the most painful of these revelations. Dave, for instance, later told the court he hadn't been expecting Brendan's visit, that he'd understood the chisels to be a gift of sorts, made two years earlier, by someone he really just tolerated.

Jane Dixon, cross-examining Kaz, spoke gently. 'Generally, how did Brendan perform at school from the very beginning?'

'Oh, pretty poorly.'

'And it was you and your wife's belief that it was some sort of birth-related brain issue?'

'Yes,' Kaz answered, 'yes.'

'And it is only recently that you've become aware that your son has been diagnosed with autism?'

'That's correct.'

No one spoke of such a condition in the 1970s; the Sokaluks had never heard of it when Brendan was a kid. He'd done badly at school then, before the Monash position, had a series of jobs, one at a rose farm near the Hazelwood pondage, and others with landscapers.

'[H]ow did those jobs go on the whole?' Dixon now asked.

'Not very well.'

'Would he last very long?'

'No. They, you know, they just got rid of him.'

Dixon asked Kaz to have a look at some photographs of Brendan's house taken by the police, and a document camera blew them up so the court could see them.

'Lounge room,' Kaz said of the first photograph.

On a coffee table in the pink and green room, a television guide lies next to the remote control, ready for the cartoons Brendan liked to watch. The room isn't dirty; Kaz has kept the floors swept, the surfaces clean, but almost every object looks forlorn, including the yellowed pillows and duvet on the old couch where his son had lain to take in his scheduled television.

'Next one?'

'That's still part of the lounge room, but it goes into the hallway.'

There are touches here of Lou, a print of bright parrots on the wall.

'Yes?' prompted Dixon, showing another photo.

'That's the bathroom and toilet.'

The walls here are the colour of fairy floss. The toilet has a fluffy pink toilet seat cover and a matching foot mat, dirtied over the years by Brendan's boots. A can of air freshener is perched nearby.

'Yes?' asks Dixon, showing the next photo.

'That was the, like a dining room beside the kitchen,' Kaz said.

A small, cluttered space with another television and guide. Amongst the clutter is a bottle Brendan has marked *Brocky's Money Box*.

'Yes?'

'And that was the kitchen.'

There's a single glass waiting by the sink. One plate in the dish rack, alongside a knife and fork.

A photo of the outside of Brendan's modest brick house was blown up on the screen. Despite its utter lack of adornment, the house isn't that different to the others in the street, but now it's the saddest. Kaz had thought they had him settled there, this different child of his, who he misses. He thought this house would make Brendan safe.

'Thank you?' The barrister urged him along.

Kaz had been Brendan's keeper, but although he'd done his best he hadn't managed to keep his son out of trouble.

'That's the house in Sheoke Grove . . . as it was on the day of 12 February 2009 when he was arrested?'

'Yes, yes.'

Those in the sealed splendour of the ornate courtroom all now saw the carport full of auto parts, and a trailer containing whatever rubbish Brendan had last hauled back – some plastic chairs, a broken clothes rack. They saw various images of the inside of his shed. Handlebars, bumper-bars, electrical cords, cookware, all piled together and looking as if they'd melded into a single mutant thing. The *DSM-5* lists 'a strong preoccupation with unusual objects' as a characteristic of autism.

'It was a mess,' Kaz admitted of the hoard, 'all the bits and pieces he had.'

The closest Brendan got to other people, it seemed, was handling their cast-offs: the mattresses on which they'd once lain, the whitegoods that had washed the baby clothes, plastic tricycles and bikes from Santa, now dumped. All tokens of the connections that eluded him. Temple Grandin memorably described herself as akin to 'an anthropologist on Mars', piecing together the rules and customs of the society into which she was born. And if you want to decode human mores, why not analyse their leavings? Brendan dealt with the scraps of other people's lives, the *stuff* that once helped hold things together. Maybe, by finding use in the useless, value in the insignificant, he was rehabilitating objects that were totems of himself.

Kaz stared at the junk.

'As far as you were concerned,' Dixon asked, 'it was just [a] Saturday morning like any other Saturday morning?'

'Yes.'

'And did Brendan's life pretty much run by routines?'

'That's correct.'

Everyone in the courtroom was no doubt wishing those days had held, that the two of them had kept going with this life of discount car shops, and long lay-bys, losing on the horses, and KFC, and that Brendan had just stayed home that Saturday as his father had told him to, rather than venturing out into the heat, into the woods.

Over the next few days in court, there was a strange re-enactment of what followed after Brendan left Kaz. A slow-building tumult of faces and voices and fragments of the story. There was the store manager of the petrol station where Brendan had bought the packet of cigarettes at 1.16 pm. They'd chatted a little. The manager lived with Brendan's brother's ex-wife. Then came a procession of locals from around the Valley, who fifteen minutes afterwards had suddenly noticed two plumes of smoke rising on the horizon.

Members of the Churchill fire brigade had been waiting on a day 'too hot to go out and do anything . . . the hottest [they'd] ever seen it'. One volunteer told the court they'd rushed towards the hills, following this 'smoke still whitish grey . . . starting to lift quite well into the sky . . . up in the clear sky above all the trees'. At the plantation raged two parallel fires: 'we seen both sides of the road were ablaze' with a 'high north, hot wind . . . forcing the fires deeper into the bush'. The radiant heat was already so extreme, one volunteer said, 'I could have cooked an egg on my face, I reckon.'

The residents of Glendonald Road heard a volunteer on this truck calling, 'Get the fuck out, quick!' A woman glanced behind her and realised that 'the whole hillside was just ablaze and there was just flames shooting above the highest trees . . . an unbeliev-able sight . . . the tallest trees, it was way above them'. Rushing to evacuate, many people noticed a light-blue car parked askew. Then Natalie Turner and her boyfriend, having heard helicopters and seen 'the flame height above the canopy of trees', picked up Brendan, who seemed 'a bit sketchy or a bit simple', and drove him home.

There was a firefighter who later in the afternoon saw Brendan back in the fire zone walking his dog. He and his crew soon became entrapped and had to fight for their lives outside the house of Geoffrey Wright, who himself told the court of the radiant heat forcing him to release his family's pet horse. ('It hated my guts, I hated its guts, but I didn't want nothing to happen to the horse because I didn't want my family getting upset.') At the firestorm's climax, the crew came to the door, burst in and collapsed. 'I've been affected by this fire immensely,' one of them said in evidence, 'and I will never [volunteer] again, never.'

There was the helicopter pilot who had water-bombed the fire, before the pyrocumulus cloud became too dangerous to navigate. And Peter Townsend, who had worked with Brendan at Monash, thinking this would be his last day: 'The fire wouldn't go out. It was up in the trees, and it was burning and it wouldn't die. It wouldn't die.' The Fergusons, Townsend's neighbours, told the court they were trying to save their house, and their lives, when suddenly they found Brendan in their midst with his dog, seeming 'very calm . . . or very vague'.

These witnesses evoked the sound and the heat and the raw terror of the fire. They were country people with few airs and with blunt turns of phrase. Many were clearly still amazed they'd survived, and some of them had not entirely recovered. Their neighbourhood was different now. Doors weren't quite as open, people were more conscious of strangers. Some days it could feel like the fire was not completely out.

On the eleventh day of the trial, Faye Last, who organised the distribution of the local newspaper, the *Latrobe Valley Express*, gave evidence. She had known Brendan since he was a small boy in the early days of Churchill. After he left Monash she gave him a delivery job, and he was a good worker. He carried his stack of newspapers in a converted pram, perching Brocky on top, before updating to a furniture removalist's trolley.

On the Tuesday after the fire, Brendan arrived on a pushbike, with the dog in a backpack, and told Faye about his car being destroyed. He'd come to collect his pay. He wanted to buy shampoo to bathe Brocky; the poor creature, after all, had been in the middle of the fire zone with him.

'What if the fire had come close?' she'd asked, regarding the dog.

'Well, it can run,' Brendan answered.

Faye's son, Jacob, had been on that first Churchill fire truck warning people along Glendonald Road to evacuate. Brendan now complained to Faye that if these volunteers had just towed his car 500 metres down the road, it would have been saved.

Faye told him the firefighters probably had more to worry about at that moment than his car.

Brendan then said he'd seen a suspicious DSE worker up near the plantation in a white ute. He told Faye he had 'first dibs' on whoever had lit the fire because he'd lost his car, and reckoned he didn't have enough insurance to get a decent one to replace it. 'I don't want a girl's car', he said. He also told her, 'I got second dibs on them for losing my watering hole,' meaning where he caught crays. 'And third dibs for polluting the air I breathe.'

Brendan took his pay, and with the little dog still strapped on his back, rode off.

Charles Szulc had lived in Churchill next door to Brendan for a few years. On the Sunday morning after the fire, he heard a knock on his door, and answering it found his neighbour, still smelling of smoke. Brendan told him his car had broken down on Glendonald Road and was now incinerated.

'I was up there,' Brendan said, 'helping fight a fire.'

The next day, Szulc told the court, he was washing ash off his car when he saw Brendan and called him over. 'You're lucky to be alive,' he said, 'because that's where the fire started.'

Brendan took this the wrong way. According to Szulc, 'he thought I was having a bit of a go at him because he sort of went back and he says, "Oh, look, someone's already accused me of lighting the fire" . . . some old bloke . . . in grief because he lost his house or something . . . Then Brendan said, "If you saw what I saw," and I said, "Well, what did you see?" and you know, he stopped, you know, and he says, "No, I can't tell." I said, "What did you see?"'

Szulc seemed unaware that the two of them could have been playing out a plotline from a soap opera, thrust from extras into major roles.

'I said, "If you know anything tell me," because I had people I knew, you know, their relatives had died in the fire, and everything like that . . . And eventually, you know, I kept on quizzing him, and he said he saw this DSE ute drive across in front of him or something, and he reckons they started the fire or something . . . And I said, "You've got to go straight to the police!" And he said, "No, I'm not going to the police . . . they twist things round." So, he sort of walked off and I left it at that, and about ten minutes later he came back . . . and he said, "I've got to go down and make a police report and I'll tell them then." So from that I took it he was going tell them, and then Thursday he got picked up, so I naturally thought he got in touch with them.'

Juries don't usually feel comfortable glancing over at the accused in the dock, but in this case, Jarrod Williams noticed, they couldn't resist. Their gaze kept slipping back to the man at the centre of this story who sat drawing intently with coloured pens, or else staring into space, looking entirely vacant. When he picked up a small plastic water cup it was with both hands, and he'd give an audible *ahh* after taking a sip.

On the trial's thirteenth day, Ross Pridgeon, the first wildfire investigator on the scene, was called to give evidence. He had also appeared before the royal commission regarding his role in the DSE's mapping of the Churchill fire, and the lack of public warnings about its speed and direction. He had found this process incredibly stressful, and discovered around the same time that he had cancer.

Led by Elston, Pridgeon now went through his qualifications. He had a degree in forest science, and during his 27-year career at the DSE he'd worked in all aspects of firefighting, fire management, fire modelling – or predictive analysis – and fire investigation.

On the morning of Sunday 8 February, having been given an approximate grid reference for the Churchill fire's area of origin, Pridgeon drove to Glendonald Road, clearing the police block. The coordinate took him to a blackened site affected by high-density fire. The burn marks around the actual area of origin would have to be lower and more haphazard. Now, from the witness stand, he gave a lesson on fire indicators: by following the signs of leaf freeze, soot levels and scorch patterns, he'd made his way to the area of confusion, the place where the indicators pointed in different directions. He was near the origin. Locating and examining this first scene had taken nearly two and a half hours.

When Pridgeon was subsequently told by a local police officer that there were suspicions of a second deliberately lit fire, a hundred metres away on the other side of Jellef's Outlet, he hunted for and found another ignition site, following the same procedure.

He told the court that after the Arson Squad's forensic chemist, George Xydias, arrived at 2.50 pm they'd together looked over each site, searching for any signs that would rule out arson. They encountered the burnt remains of a fair amount of rubbish, but nothing that would self-ignite.

Three times a year HVP, the company that owned the plantation, hired a contractor with an excavator and a tip truck to remove junk – old fridges, furniture, even cars. This public dumping was not much better in the national parks Pridgeon helped manage. He had got into the job through a deep fascination with nature, and it could fill him with quiet fury that people thought of the forests as mere refuse stations. They liked a nice view for a picnic, but entirely missed the point that their own future was intricately connected to the forests' health. That was what no one seemed to have learnt from Black Saturday. Fire science wasn't some obscure area of academia, it was intrinsic to our understanding of the country and our safety within it.

If Pridgeon now found that his irritation gave him some energy to assert himself in the courtroom, despite his nerves and precarious health, well, that was useful in the testy exchanges that were to follow. When he was cross-examined, it was immediately combative.

In his report on the Churchill fire, the defence's wildfire expert, Dr Kevin Tolhurst, had lambasted Pridgeon and Xydias's investigation.

'You would regard Tolhurst as a well-respected expert in the field of forest fire science,' Dixon suggested, after repeatedly questioning Pridgeon's qualifications, 'and you are aware that he's published a great many papers about wildfire and forest fire science?'

'Yes.'

'And lectured in the area?'

'Yes.'

'And gave evidence on several occasions at the Bushfire Royal Commission?'

'Yes, I'm aware.'

The judge interjected and reminded Dixon that many people gave evidence at the commission.

Dixon, who had a flinty demeanour in court, continued: 'You have read [Tolhurst's] report and you understand that he has raised a significant concern about the absence of any source of ignition being found in this case?'

' Yes,' Pridgeon answered. 'I have read that.'

'And he, in a nutshell, puts forward the position that in the absence of finding such a source of ignition, the chances of solving the cause of the fire are reduced?'

'Not finding a causal agent doesn't reduce the probability of arson.'

'He indicates that one of the purposes of wildfire investigation is to try and find the actual point of origin?'

'That's always the aim, yes.'

'Or source of ignition. And that was an aim that wasn't able to be met in this case?'

'Correct,' Pridgeon replied.

'And so in a sense the area of origin wasn't particularly well defined?'

'I believe it's quite a good area of origin for that type of fire . . . [T]he combined group of the Arson Squad and myself thoroughly investigated that area, looking for any evidence that was possibly there.'

Dixon started to line Pridgeon up for the defence expert's greater criticism. 'You are aware from reading Dr Tolhurst's report that he identified a separate area of origin that he's called Churchill Two?'

'Yes. I'm aware of that.'

But what Pridgeon knew and Jane Dixon was about to discover – disastrously for Brendan Sokaluk's defence – was that Kevin Tolhurst had made a grave blunder. The defence barristers slowly had it revealed to them that this separate fire the police investigators had supposedly ignored through sheer incompetence was non-existent.

Detectives Bertoncello and Shoesmith had spent hundreds of unpaid hours finishing off the investigation of this case to their satisfaction. They had requested from the prosecutor a copy of Tolhurst's report and had instantly felt it contained significant errors. By closely examining the photographic evidence taken by the public on Black Saturday, they had proved that Churchill Two was actually the result of a spot fire from the main blaze. It had started to burn in an adjacent valley hours after that main fire had.

Ray Elston, using this evidence in his re-examination of Pridgeon, now showed the jury in excruciating detail that Tolhurst was mistaken.

George Xydias took the stand after Ross Pridgeon. In his late forties, he was a slight man with deep brown eyes and thinning hair. He couldn't count the number of times he'd been to court during his nearly twenty-five years of working as a chemist with the Arson Squad, but the scientist was an introvert by nature and regardless of how big or small the case was, giving evidence always made him nervous. Xydias had investigated more than a hundred wildfires himself – an increasing number of them lit by serial arsonists – and attended countless others to assist or train less experienced investigators.

Black Saturday had perhaps exhausted him even more than it had Pridgeon. Over an eighteen-month period, Xydias had either initially examined or reviewed the examination of all 173 fatalities. After attending the areas of origin near Churchill, and then the connected crime scenes where people perished, he and his assistant had rushed to Marysville, where the Murrindindi blaze had nearly wiped out the town and killed forty people. One night, a coughing Xydias had tried to sleep in a room full of soot and smoke with the fires still raging 300 metres away. From Marysville he moved to Kinglake, which had the same number of casualties and level of destruction.

Xydias knew more about combustion than he wanted to. In his career he'd seen literally thousands of bodies in all states of devastation by fire. The police psychological services didn't often reach out to the chemists, but when they did, Xydias wasn't tempted into counselling. He thought it would be disingenuous to say that his job hadn't affected him – in fact, far from being inured to death, he couldn't bear to look inside an open coffin – but he didn't believe that someone who hadn't seen what he'd seen could advise him on how to deal with it. His means of coping was to move on from each job quickly. He had to, anyway – there was always another. Each new traumatic scene was likely to remind him of an earlier one. Children and pets got to him the most – those with the least agency, who hadn't made the choice to stay.

In the courtroom, a projected map showed the intersection of Glendonald Road and Jellef's Outlet. Elston asked Xydias to indicate the fire's two areas of origin. The one on the eastern side of the outlet began three to four metres inside the plantation, and Xydias narrowed the total area down to eight square metres. On the western side, four metres into the plantation, he could get the area down to five square metres.

'Did you make observations as to whether there were any electrical or powerlines in that area?' Elston asked.

'No, there was nothing in the area,' Xydias said.

'Were there any transmission towers of any kind in that area?'

'No.'

'A discarded or improperly extinguished cigarette or cigarette butt – what do you say as to that being the cause of fire in the area?'

'With a situation like the day that we had – severe heat, wind and the conditions of the terrain,' Xydias explained, 'if there was only one point of origin identified, I don't think we can exclude a cigarette being involved. The problem we have at this particular scene is that there were two distinct and separate areas of burning. They were unexpected in the sense that one could not have come from the other. They had to be distinct.'

'Did you come to any conclusions about the source of ignition that was required?'

'I assumed, being deliberate, that it would have been something like a lighted match or a cigarette lighter.'

When Dixon began her cross-examination, she countered, 'There's an element of assumption about all of that, isn't there?'

'Yes,' said Xydias.

Dixon was now in a difficult position, however. The fact that there was no Churchill Two fire quashed the idea that the hills were crawling with arsonists on Black Saturday, and made her attempts to discredit the original fire investigation a riskier tactic. She nevertheless suggested that Xydias had relied too heavily on Pridgeon's findings, that his qualifications weren't up to date, that his notes on the case had been too brief, that he hadn't looked closely enough for an ignition device, and that he'd too hastily discounted Tolhurst's other significant theory, that the fire on the eastern side of Jellef's Outlet was actually the result of a spot fire from the fire to the west.

Xydias, who held Tolhurst's report in low regard, methodically discounted each assertion.

Dixon said, 'Essentially your position is: Well, I consider there's two different areas of origin, I therefore think a cigarette as an explanation is unlikely. That's it in a nutshell isn't it?'

'Well, I'm saying two cigarettes at roughly the same time is unlikely, yes.'

In his re-examination, Elston asked at what point in time Xydias would expect this fire to have started creating spot fires.

'Well, by the time this gets to a stage where it's large enough to spot, I . . . would say it would be in excess of fifteen, twenty minutes. I can't see it happening significantly before then.'

Xydias didn't glance over at Brendan Sokaluk in the dock. He made it a point to never look at the accused. When the scientist walked out of the courtroom, he hoped this would be the last time he had to think about the case. Thinking about it made him remember.

Adam Shoesmith was no longer head of the Arson Squad. After Operation Winston, he'd been moved to the Purana Taskforce to investigate what were known as the gangland killings, a series of crime-world assassinations as different groups vied for control of Melbourne's illegal spoils. From there he'd been attached to the Australian Federal Police, tracking major international crime syndicates running large-scale drug importation schemes.

Now he climbed the steps to the old-fashioned witness stand, and in this courtroom with such high ceilings felt as though he were perched in a crow's nest, or, worse, a gibbet – one of the hanging cages he'd seen jutting off the walls of eighteenth-century castles. The room was cold, with terrible acoustics. Shoesmith had poor hearing and needed to concentrate.

'On Tuesday 10 February 2009,' said the prosecutor, 'were you appointed to manage the investigation of what is now known as the Churchill Fire?'

'Yes.'

'After briefings, you attended the Morwell Police Complex, is that so?'

'I did, on 11 February.'

This back-and-forth established the sequence of events that

led to Brendan's arrest on 12 February, and then what looked to be a 1980s-style projection screen was wheeled into the courtroom, and a silent, sixteen-minute video was played. It showed footage of Brendan's house, images the defence had tried to soften earlier by having Kaz introduce them. Each room looked bleak, shabby. In the bedroom an old mattress was decked in filthy sheets and pillows; a great teddy bear lay on a wardrobe shelf amid crumpled camouflage-print clothes.

Shoesmith's work had made him a professional cynic and he still wondered if Brendan was as impaired as his lawyers claimed, but watching this footage in court he could see that the house emanated unhappiness.

In the blush-coloured room, by Sokaluk's computer, was a notepad with *Churchill Police* and the telephone number written down. Was he planning on calling to report the DSE arsonist, as he'd told his neighbour? At the bottom of the pad, Sokaluk had written his surname and Alexandra's surname, surrounded with doodled embellishments.

On another piece of paper were jottings arranged as a list. It read like a poem on loneliness:

11-10-69 [Brendan's birthdate]
green scrape
yes. Alone
wife. Kids
to be happy with you
people who bitch a lot
not yet.

When the video had finished and the court's lights were raised, Elston said, 'Now, as you told us, Detective, you went back to the Morwell police complex at about ten past six?'

'That's right, yes.'

Shoesmith explained that he'd organised for the forensic doctor to make the two-hour trip from Melbourne to assess Brendan's mental state. The accused man was lodged in the cells and given dinner. At around 8.40 pm, Shoesmith was told that Brendan had asked to see his colleagues, detectives Henry and Bertoncello.

A transcript of their subsequent interview was given out to the jurors. The judge, taking pity on still vertical Shoesmith in the witness stand, told him he could sit. His best option was a seat close to Brendan in the dock.

Next the police interview started playing on the old projection screen. In the dim light, the Brendan of three years ago, in his olive-green shorts and sweatshirt, was blown up: '. . . I burnt down one thing I loved in my whole life is the forest and my stupid actions stuffed it up. Now I have no place up the forest, going to sit and watch the fish, look at the creeks . . .'

Out of the corner of his eye, Shoesmith could see Brendan drawing something, seemingly uninterested even in this apparition of his younger self: '. . . my life gets hard from the stuff, and stress wise and that, I would go up there and sit, and this would to relax, I found. But now I've destroyed all those areas and all those poor people died, so stupid.'

This was the first time the jury had heard the drone of Brendan's voice. Shoesmith had grown used to it. He'd been listening to the man's prison phone calls on and off for the past few years. He'd heard Brendan discussing with Kaz the price of scrap metal, and asking after Brocky, and slagging off about the police being pigs. All the time, Shoesmith had wondered if, in the midst of those stop-start cadences, Sokaluk would slip and admit to deliberately lighting the fires.

The court paused the video and Shoesmith returned to the witness stand. He gave evidence that on the night of Brendan's arrest, the digital imprint of a Crime Stoppers form had been found on his computer. He'd submitted a report claiming that a DSE firefighter had lit the fire.

Shoesmith sat back down again near Brendan. The accused was strangely more life-like enlarged on the screen. In the police interview, he admits: 'I sent a thing . . . so people wouldn't blame me and then they wouldn't hurt me.'

The footage continued with Brendan, the following day, leading the detectives to the intersection of Glendonald Road and Jellef's Outlet. Standing amongst the trees, which he noted were 'all charcoal', pushing his hands in and out of his pockets, he showed Adam Henry where he'd dropped the serviette-wrapped ash. Within minutes, he said, he realised a fire was burning.

'How big was the fire when you first saw it?' Henry asked.

Sokaluk's gaze shifted. 'It was big.'

'When you say big, describe it?'

'Too big for me to put out.'

'All right, well, size-wise,' Henry tried, 'would you compare it to a footy field?'

A beat. 'No, probably a couple of car sizes . . .'

The blackened gums might as well have morphed into the bars of a cell. Before long, on the screen, Brendan was denying he'd deliberately lit the fires, 'No!' . . . No! I had no intention for all this to happen. Now I've got to put up with it for the rest of my life and it makes me sad.'

But Shoesmith was only half watching this. He was preparing himself to be cross-examined. He knew Jane Dixon planned to run a line on alternative suspects, and 80 per cent of the work he'd done on this case was to eliminate such a defence. The Arson Squad had spent months checking every other potential suspect. The nomination of some candidates had been ridiculous – people reported by whoever disliked them – others took slightly longer to rule out. Shoesmith readied himself to recall the details of each.

'In terms of the investigation of this fire,' the defence barrister started, 'there were a number of sightings of motorbikes on the day, and around the general time period of the fire, that were reported and investigated by police?'

'Yes,' Shoesmith said. Much of the potency had gone from the insinuation of a 'more mobile arsonist', now that the defence couldn't prove other random ignition sites. The detective felt he wasn't going to be blindsided.

A to-and-fro ensued, Dixon listing sightings of various motorcycles ('a man apparently in overalls, watching the fire through binoculars or a camera?'; 'a motorcycle riding around . . . with what appeared to be a jerry can or something strapped to the back?'), and Shoesmith confirming times ('2.21 pm, fifty minutes after the fire's start'; '4.30 pm . . . with a container, not necessarily a petrol container on the rear of it').

He could feel the tension building in the courtroom. There were only two days of evidence left, and this trial was nearly over. But he had learnt not to try to divine what a jury was thinking; and who knew whether, when the defence's well-known fire expert appeared, they'd believe what he had to say.

Shoesmith's mother had passed away a few months earlier. While she was in palliative care, he'd worked on a submission opposing Brendan's bail application. Summer was approaching and it was his belief that the accused was a serial firelighter. He'd studied the CFA records from the time Brendan had been a volunteer, and thought he was likely responsible for a number of the blazes at which he'd appeared without any notification. And in more recent years, at the ignition sites of other fires lit around Glendonald Road, remnants of paper serviettes had been found in the debris.

The night the detective's mother died, he sat with her body until she was taken away. Then in the morning, he appeared at the bail hearing. The lawyers, he felt, had gloated when the ruling went their way. Shoesmith knew it was illogical, but Sokaluk's conviction was something he now wanted for his mother. (It was easier, too, to get angry about Tolhurst's claims of a non-existent fire than it was to grieve.) Shoesmith usually tried not to get emotionally involved with victims' next of kin, but this case had gone on for years and he'd got to know the families. When his mother died he felt, somehow, in his mourning, that he joined their ranks.

As the trial wound to its close, the defence badly needed something to go their way. One of their witnesses, a materials engineer, had theorised about hot embers exiting the long exhaust pipe of Brendan's 1974 HJ Holden, setting the bush alight. So too could a 'meteorite', the judge quipped without the jury present.

Brendan kept drawing. It was doubtful he got the joke.

Another defence witness, Dr Marged Goode, had told the court that in her practice as a psychologist she'd found many people on the autism spectrum preferred not to have to be alongside others day in, day out, adding that 'given that we are group animals, not being able to relate to those in our group' is extremely stressful. It was no wonder, she believed, that the children's television shows Brendan favoured had soothingly repetitive plots, with no nuance or subtext. Thomas the Tank Engine's co-workers, the other Sodor trains, said only what they meant. They shared no sly, ambiguous asides, and their simplified faces conveyed clear emotional information.

In any given episode of the cartoon, there was usually some unfortunate mix-up and then, within the ten-minute format, a swift resolution and forgiveness.

These days in court, on the other hand, had stretched on and on, with the judge in his Santa suit listening carefully and then 'blabbering on about stuff', as Brendan told his lawyers, which he found weird and boring.

In the dock, Brendan was now repeatedly sketching with his red pen, then his blue pen, the DSE ute he claimed to have seen on Black Saturday. This vehicle, emerging over and over in his notebook, seemed to have become more real to him as the facts of the case made the dropped cigarette story less plausible. The defence lawyers saw these pictures and felt dismayed at the ground shifting underneath him.

Dr Kevin Tolhurst was a friendly man with a bushy red beard and glasses. The last witness to appear, he took the stand on the trial's eighteenth day, and Jane Dixon recited his extraordinarily long and impressive list of fire-related credentials, scientific and academic.

Amongst his achievements, Tolhurst was regarded as a leader in the field of predictive modelling, which sought to map a fire's behaviour and likely spread. The charges against the man accused of lighting the Delburn fires had been dropped after the scientist convinced the judge at the committal hearing that, despite this suspect being seen in various compromising situations and locations, he was, based on the modelling, innocent.

Here, too, Tolhurst was appearing for the defence. But the judge in this case was less impressed. In discussions without the jury present, Judge Coghlan had previously told Dixon that Tolhurst struck him as 'barracking', 'standing on top of the hill throwing rocks down at the combatants below'. His was 'the worst kind of report', Coghlan believed, in taking a side and stalwartly attacking the initial fire investigators, 'and I have a view about who I think it hurts most . . . I mean is this jury seriously going to accept that Mr Xydias didn't look for the source of the fire because some smart fellow says, years after the event, he didn't?'

Giving his evidence, Dr Tolhurst reiterated his belief that a spot fire was responsible for the eastern ignition point found on Glendonald Road. But Tolhurst's cross-examination was sport for the prosecutor.

'If the winds have picked it up and it's running south-east pretty strongly,' Elston said, 'it's a peculiar dynamic, isn't it, that makes it blow to its left and backwards?'

'Well, I've been to many fires and see it all the time,' Tolhurst replied. 'It's not that peculiar. It happens.'

Unfortunately for Tolhurst, even if this was a dynamic he'd seen before, his gaffe over the mapping of Churchill Two meant his credibility had taken a fair blow.

'The reason you posited the existence of Churchill Two was to create the situation where there might have been somebody else running around starting fires in there, wasn't it?'

'Well, it raised that possibility, yes,' Tolhurst admitted.

In a further embarrassment for the witness, Elston took out a protractor and now proved that on one of Tolhurst's maps, north had been deleted and replaced, pointing in a direction that was fifteen to twenty degrees out, skewing the angle of the wind and increasing the likelihood of the second fire being naturally lit.

'It's indicative of north,' Tolhurst said of his map, 'it's not meant to be used for surveying or navigation.'

'Okay, it's pretty significant when you start talking about directions of winds, fire, as to where the cardinal points of the compass are, isn't it?'

'Well, we are only talking about north, south, east and west.'

'North is north, isn't it?' asked Elston. 'If you are going to go somewhere to the north?'

'No magnetic north is true north . . .'

'You are going to tell me you went playing with magnetic north?'

'No, I was just showing the approximate direction of north,' Tolhurst quibbled.

Earlier, the prosecution fire experts had both given evidence about the difficulty of actually starting a fire with a dropped cigarette. The cigarette would need to land on, and embed in, very fine types of forest fuel and be attended by just the right airflow, and even then it would still only occasionally produce flames. The iconic 'cigarette out the car window' – a classic Australian story of carelessness begetting fire – was highly unlikely to create a bushfire.

Elston wanted to know what Tolhurst thought the chance was of a fire starting in the manner Sokaluk had described: 'an ember squished in a napkin, flicked out the window'.

'If it didn't have any unburnt tobacco going with it, yes, I would say that it's a very low probability of starting a fire,' Dr Tolhurst conceded.

All this while, Brendan sat busily conjuring the wheels and body and windows of the white ute, perhaps wishing it to life so he could drive out of there.

In her closing address, Jane Dixon took the jury back again to the moment of the fire's birth. There, in the crazing heat with the eucalypts avid for a spark, the first flames stretched in every direction, a picture of havoc. 'The case before you,' she said, '[is] a bit like the area of confusion which was described by the fire experts . . . an area where there are different indicators pointing in different directions. The flags don't all point the same way, and that's how you know you're in the right spot, apparently. And we say that's a bit like this case . . . The flags don't all point the same way, in the direction of guilt[.]'

In the dock was her confounding client. He had been right

there when the plantation's fire was a 'couple of car sizes', and had called 000, in the defence's telling, as a concerned citizen, 'hardly the act of an arsonist'. As his trial reached its end, Brendan Sokaluk looked less like someone whose fate lay in the balance and more like someone waiting for a bus. His expression remained remote, he yawned, he gazed into the middle distance. When the jury stole glances at him, he surely reminded them of *that person*, whose otherness they might see and instinctively avoid, and then feel guilty about the impulse. His presence in this courtroom evoked both pity and outrage, and that was the greater area of confusion.

The defence knew the momentum of the trial had not gone their way.

'You weren't lucky,' the judge told Dixon during a discussion before her final address. The defence's case relied on expert evidence, and their witnesses hadn't struck the judge as 'very expert'.

'North is always north as far as I'm aware,' Ray Elston had scoffed while summing up the prosecution's case. In his final address, he'd claimed that the direction of the fierce wind on 7 February 2009, a northerly, eliminated the possibility of the two adjacent fires being anything other than deliberately lit. And lit by this man, who'd led detectives to a site metres from where the fire investigators had located the signs of ignition. Sokaluk, although impaired by autism in communication and social skills, was still able to set a firestorm then lie about his involvement.

With the evidence tightening around Brendan, Dixon argued: 'This was no contrived web of lies and deceit . . . this person is a simple man who seems to have believed himself in very, very big trouble.' She urged the jurors to consider her client as he appeared in the tape of his police interview, without 'any prism of prejudgement'. His 'patchy, uneven neurodevelopmental disorder makes him a poor judge of his own thoughts and actions and place in the world'. But by looking carefully they would see his distress over the fire, and his repeated and persuasive denial of guilt: *No! No!* 'It's stark and it's plain and concrete and real. And what Brendan does is tell the truth as he perceives it to be.'

All along it had been difficult to entirely believe Brendan's explanation: the fiery interplay of cigarette ash and a napkin. And now Dixon was conceding to the court that it may well have been an incomplete version of events. But Brendan's grip on reality was fundamentally skewed. 'He knows there's a fire, he thinks, "Oh, oh! Did I cause this? That thing I chucked out, that serviette with the burning cigarette fuel, did that cause the fire? . . . It mustn't have been properly out. It's all my fault. My fault as usual." . . . It's *not* some cunning made-up story.'

But it was a story nonetheless. Whether it was one invented out of slyness, to disguise a criminal act, or naivety, because Brendan had trouble perceiving and remembering his fumbling actions, was the central conundrum and point of dispute.

Dixon reminded the jury that Dr Goode had helped paint a portrait of a man who had been mistreated for much of his life, both by those who'd actively wished him harm and those who simply hadn't understood his distinct psychology. Hounding Brendan to the point of meltdown became a sport for his class-mates. The child who relied on sameness in order to cope was punished as he floundered. Hiding in a storeroom when his regular primary school teacher was away, possibly having a panic attack, he was forcibly dragged back to the class by the other staff. His school reports stated that when he was struggling he didn't ask for help – but he hadn't conceived that other minds might have a solution. At work, with a focus on different stimuli, he had great difficulty following instructions, and with a poor feel for his tools he was more injury-prone. He was treated, though, as a pest and a malingerer: the 'half-wit' whipping boy.

It was a hard irony that, had Brendan been better able to process verbal information, some of the experts' evidence might have helped him understand his difficulties, and possibly provided some solace. The answers to what Goode claimed were standard questions for people with Brendan's disorder: Why don't people like me? Why am I different? Without this piece of the puzzle, Brendan – and his schoolmates, and then his workmates – entered into a common downward spiral.

As Goode put it, 'We all tend to react in a negative way to people who don't behave as we expect. It's not necessarily a conscious process but the interaction has been skewed already.' The autistic person, habitually treated as weird, or retarded, has 'great difficulty determining whether hurt is deliberate or accidental. So, they kind of get into the habit of assuming that it's deliberate . . . they get defensive, they may retreat, they may become sullen, they may do all sorts of things which make them appear even worse to somebody.'

The questions hovering unanswered in the courtroom were: Did Brendan accept responsibility for the fire because he was so used to being the one in trouble? Or did he light the fire to revenge himself for his poor treatment? Or did the truth lie somewhere in between?

The defence lawyers had found their client to be a compliant, almost docile man, but they knew from the statements in the brief of evidence that this was not everyone's experience – particularly at Monash, where Brendan had mixed childish stunts with behaviour that his colleagues found belligerent and spiteful. A female apprentice quit because Brendan would sit staring at her, or make grunting noises, or block her path, making the job intolerable.

'Whenever Brendan was upset there would invariably be repercussions,' a co-worker had told the Arson Squad detectives. Nothing could be proved, but equipment would be found damaged, tools missing, 'and more personal things like . . . I had the wheel nuts undone on one of my wheels'.

'He wouldn't relate properly to people and it seemed to me that he would deliberately do something to push people's buttons,' another colleague reported. 'It was as if he got off on making you angry . . . I was very scared of Brendan. I didn't know how to deal with working with him because Brendan would often get pissed off at something that you would do and then make threats to blow you up and kill your family.' The colleague reckoned Brendan said he could get dynamite from relatives who owned a quarry. 'He would sometimes smile when he said it, but I never knew whether he was joking or not. I found it very stressful. I reported it to my supervisor several times, but I was told to just deal with it.'

Brendan had apparently told this supervisor, 'I know where you live,' and when he answered, 'I live a fair way off the road,' Brendan answered, 'I could still shoot you with a rifle.'

The defence lawyers had interpreted these anecdotes about his aggression as the acts of a cornered child striking out, making outlandish claims borrowed from his cartoons. A small boy might threaten to get dynamite and blow away his enemies – coming from an adult it sounded ludicrous. These colleagues, however, had found him unpredictable enough to still feel chilled.

Brendan had spent a lifetime trying to pass, to find ways of appearing less impaired to those around him, and one effective method was through intimidation. But autism is by definition asocial rather than antisocial. To be antisocial, at least in terms the *DSM* recognises, a person needs to grasp basic social concepts. They need a competent working knowledge of how other people operate. They understand when they're causing harm to others, and either don't care – perhaps the better scenario – or enjoy it. Whichever, people with antisocial personality disorder are attuned to the nuances of their victims' feelings, and to the suffering. They have what psychiatrists refer to as a deficit in emotional empathy.

It would seem that instead Brendan Sokaluk had a deficit in cognitive empathy. In his blunted way, he was unable to understand his colleagues' feelings, to take their perspective. For years the consensus was that those with autism felt no empathy, and that, being unable to understand another's grief, they were therefore incapable of remorse. But those on the spectrum are often marked by *irregular* levels of empathy. Some autistic people are conscious of other individuals' and animals' feelings to an almost painful degree; the idea they've hurt someone can be devastating. For others (perhaps as with many in the rest of the population), empathy is a harder emotion to access.

Certainly Brendan's grasp of social situations was idiosyncratic at best, and influenced by a lifetime of being an outcast in a place where life was tough and combative.

In her memoir, *Nobody Nowhere*, Donna Williams describes her 'insular and extremely lonely, unreachable' childhood in late-1960s and early '70s Melbourne, dealing with poverty, abuse, and undiagnosed autism. 'The world seemed to be impatient, annoying, callous and unrelenting,' she writes. 'The more I became aware of the world around me, the more I became afraid. Other people were my enemies.' To them she was 'a nut, a retard, a spastic. I threw "mentals" and couldn't act normal.'

But with violence, Donna 'knew where I stood. To call it the result of "baser" emotions must be true, for I certainly found it easier to grasp. Niceness is far more subtle and confusing.' She 'began to fight back . . . trying to kick people down stairs, hitting them with chairs, slamming fingers under desk-tops and becoming hard, quiet and brooding'. Eventually, she writes, 'hatred became my only realness and, when I was not angry, I said sorry for breathing, for taking up space'.

Did Brendan feel like this?

There's an unsettling detail on the list of Sokaluk's various misdemeanours compiled, years before Black Saturday, by his supervisor at Monash. Amidst the petty complaints about him stuffing food into his pockets at work functions and mowing over rubbish, is this claim:

- At the time of the Pt Arthur massacre, Brendan boasted that he could understand the perpetrator's intentions, not aware that others would find this comment offensive.

The intellectually impaired Tasmanian man Martin Bryant, who went on a killing spree in 1996 and shot dead thirty-five people, was eighteen months older than Sokaluk. Like Brendan he had been a socially distant child who had difficulties with communication. Considered 'annoying', he was the victim of severe bullying and grew into a strange, isolated person. After his arrest, at just on twenty-nine, he was assessed as having the mental age of a ten- or eleven-year-old. Before the massacre, Bryant's hobby was taking international flights. On these long-haul trips, those in adjacent seats were unable to avoid talking to him, and later he could recount these conversations to police almost verbatim.

What was it about Bryant's intentions that Brendan could understand? What did he even imagine they were? A yearning for revenge? For notoriety? For the unleashing of a desire for power? Did Brendan feel these things too? And if you were never going to be considered a hero, was there allure in becoming a villain?

After he left Monash, Brendan spent much of his time on the internet. Whatever else he was doing online, he also played computer games. They accurately reflected how he viewed his life. On screen he was small and other, a kind of alien facing multiple threats. But in his camouflage-print clothes, he was dressed for imaginary combat, ready.

'Oh, oh! Did I cause this?' the barrister had asked, imagining herself as the accused.

What if Brendan's condition meant, as Dixon claimed, that he had lit the fire and yet remained so alienated from his own self and actions that he didn't completely grasp what he'd done? The CGI landscape changes and a fire, his devastating avatar, has appeared. The cigarette story is part confession and also a metaphor for his own lack of control of himself. An accident and a compulsion aren't that far apart. Neither is willed, both are only partially understood.

Dixon stood in the courtroom and tried to angle the fire indicators in the right direction for the defence. She urged the jurors not to feel 'under pressure to somehow make someone responsible for [this] catastrophic occurrence'; to accept the wildfire expert's thesis that the second blaze was due to a spot fire. Kevin Tolhurst testified that he'd been to countless fires and had seen 'local turbulent winds' – gusts that come back the wrong way. On Black Saturday the wind had been blustery and unpredictable; an ember could just have eddied and spiralled and started a second blaze.

In the back of Jane Dixon's mind was an inquest she had appeared at sixteen years earlier. She'd been part of the legal team representing the families of nine intellectually disabled people who died in a fire at Kew Residential Services, formerly known as Kew Idiot Ward, in 1996. It was the kind of place a child such as Brendan might have once been sent to live: a world, the inquest had revealed, of warren-like rooms with locked doors, poor staffing, and inadequate safety features. The fire had been started by a 38-year-old resident with very little language who was obsessed with matches and lighters. He would pick the staff's pockets to find them, and on making a flame appear – a tiny magic act in a dreary, loveless world – pronounce, 'Good boy!'

A flame can be an entrancing, other-worldly sight. And what Dr Goode had not told the court was that psychologists often separate autistic fire-setters from others who deliberately light fires, because some neuro-atypical people find the flames not just mesmerising, but soothing. Fire-setting can be part of a repertoire of self-stimulating behaviour used to regulate anxiety. In February 2009, the same month as Black Saturday, a Melbourne magistrate had refused to grant bail to a teenager, until he was medically assessed, who'd admitted to starting two small grassfires because 'he liked the burning pattern fires make [and] flames have infinite possible shapes'.

A fascination with light and movement is defined by autism expert Bryna Siegel as a defining characteristic of the condition. Donna Williams wrote of 'fractured patterns, which were entertaining, hypnotic and secure'. Naoki Higashida believes that 'when a colour is vivid or a shape is eye-catching then that's the detail that claims our attention, and then our hearts kind of drown in it, and we can't concentrate on anything else'.

Could Brendan Sokaluk, in his isolated world full of broken, cast-off things, have been particularly vulnerable to the enchantment of something vivid or eye-catching?

On Thursday 15 March, the twenty-third day of the trial, the jury finally retired. While Brendan's fate was being decided, the defence solicitors essentially babysat him. He watched television in a Legal Aid interview room, favouring the same programs as Shannon Dellamarta's four-year-old son. Dellamarta had worked with clients who had allegedly committed violent offences, and with whom she felt the need to stay on guard. But Brendan was never nasty or threatening. If anything, she worried for this juvenile man, knowing whom he'd meet in jail. Usually, even her intellectually disabled clients would ask how long they could be locked away for. Not Brendan, though. It didn't seem to register that he should be curious about what came next.

The three months Brendan had been on bail were happy ones, he'd told Gaby Pulczynski. He'd had his squabbles at the assisted-living facility, with some of the other troubled men there accusing him of stealing or breaking things, but that was his ongoing story. He'd also worked out how to use Melbourne's public transport system. There was so much to see, and no one knew him. He was just one face in a mass of faces, no more strange than anyone else, camouflaged by all the weird types a big city enfolds.

The jury deliberated on the Friday and the Saturday, and again on the Monday. Although the trial hadn't gone to plan for the defence, these lengthy deliberations suggested that *some*one had been convinced. Then, on Tuesday 20 March, the twenty-eighth day, the jury re-entered the courtroom. It was 11.36 am.

Detective Paul Bertoncello sat in court waiting.

Throughout these past three years, he had been the police liaison for the victims' families. When, for instance, there was a delay in the legal proceedings, he had to ring thirty different people – the parents, children, partners, siblings of the dead – and let them know what was going on. Completing these calls usually took him a couple of days. Sometimes the conversations were merely functional, at other times they were deeply emotional. Often he was managing expectations: Bertoncello would explain the difficulty of securing an arson conviction, and try to impart some general wisdom about trials. A guilty verdict, he'd warn, would not necessarily bring the pain and fury to an end.

Now he sat with two mobile phones. Only a few families had opted to come to court. Bertoncello had advised those who stayed away that he wouldn't be able to reach each of them individually before the verdict hit the media. Instead, when he heard the judgement, he'd send a group text. On one phone he'd typed a message announcing that Sokaluk had been found not guilty, on the other he was guilty. Each was ready to be sent to the thirty contact numbers.

'Madam Foreperson, apparently you have agreed upon your verdict?' Judge Coghlan enquired.

'Yes, your Honour.'

When the detective heard the single-word verdict, he felt pure relief at being able to reach for the phone holding the 'guilty' message.

Later, at Brendan's sentencing, Ray Elston had to stand in the courtroom and read the victims' impact statements. Paul Bertoncello thought the prosecutor delivered these stories in something approaching a monotone, as if adding any emotional expression to the words would break him. Even so, the detective sometimes heard Elston's voice crack.

The statements told the story of Black Saturday three years on.

Rodney Leatham, the man who had been unable to save his wife from dying in the fire, couldn't be in the sun anymore because of his burn scars, 'so itchy that I could tear my skin off . . . Every day,' Elston read out, 'I have a thousand what-ifs, that I should have done this, or I should have done that, and my wife would still be alive.' Rodney had spent five months in rehabilitation, 'but I pushed myself so I could finish, because there were others that needed the staff's attention and time. I feel I can't go places I used to go anymore because everywhere reminds me of my wife. It's as though I have two lives, the one I am trying to live and the one that is dead and gone.'

A woman whose husband and son had died found it difficult to sleep at night: the membrane was so slight between what might have been and what was. Her husband had woken that Saturday morning and fed his cows and checked his fruit trees and vegetable garden. After hearing of an approaching fire, the woman evacuated, leaving her husband to defend the property. As she drove away from the fire, a driver coming towards her flashed his lights. Her son rushing to help his father. That was the last time she saw him.

Now she was the guardian of her son's child, and this boy could hear her crying herself to sleep. 'I cry a lot and he does not like me crying, and this makes me feel guilty, because I don't want him to be sad.'

Another now-fatherless child also had trouble going to sleep, and getting up each morning, 'because I just keep thinking about my dad,' the prosecutor read. 'After Black Saturday I had to catch a bus to school and it went past my old house that got burnt, everyday.'

A woman whose brother died in the fire had lost contact with people because she didn't want to be asked, or have to answer, how she was.

Another woman worried that she would forget her brother's voice and mannerisms.

A man who'd been on the phone to his son-in-law in the moments before the latter's car caught alight asked himself constantly whether he could have done something to change what happened.

A woman who lost her husband pictured his death over and over: 'images flash inside my head of him burning every day'. She had given up her nursing job, unable to concentrate.

A woman would hear her two-year-old granddaughter calling out to her dead son, 'Uncle, where are you? Come and play with me.' The child didn't understand why the people around her were weeping, but she would wake in the night calling the young man's name.

A man who'd stood in a local fire station begging firefighters to send help to the house where his son was trapped would hear a song on the radio that his son had once listened to and be brought instantly undone.

Ray Elston went on reading. In these accounts, the bereaved told of being short-fused; of relationships breaking down between parents and children, between partners, between siblings. He read of self-harming; of needing antidepressants, and anti-anxiety and sleeping medication. Some people had had to move – if there was a house left to move from – to avoid continually driving past the scorched forest. The fire kept spreading in this other dimension, burning through memories, and the layers of identity. Aerial photographs had shown a landscape of black. The survivors found themselves still living inside it, daily tasting the ash.

The day of the verdict, Shannon Dellamarta had waited in court for the jury's decision with a pounding heart and nausea. These physical symptoms rarely struck her so severely, but the magnitude of this moment, and its consequences, suddenly felt clear. When she heard the guilty verdict she tried to stay composed. She knew the defence team's reactions were being watched. In the face of Brendan's seeming vacancy, the lawyers, and particularly she and Gaby Pulczynski, had become surrogates for the accused. They had had to absorb the grief that would wash over the relatives in the gallery, the disdain of the press. And now the police, the families, the journalists – they all wanted to see if these lawyers, 'being on his side', were suffering, how *much* they were suffering, whether they were sorry for what their client had done.

As people hugged and wept, Brendan looked as disconnected as ever.

Dellamarta knew that a perpetrator's public contrition was a flawed measure: who can really tell the depths of an offender's misery? It's so human, though, the desire to see remorse. The Latin root of this word, *remordēre*, means 'to bite back'; it's the gnawing of one's conscience. We want the wrongdoer to acknowledge the pain of the bereaved and then to feel some of it. We invented hell for the purpose.

But whether Brendan *could* take in the scale of the loss was unclear. Certainly, any remorse he may have felt could not be displayed in a way a courtroom would ever understand. Instead, to the extent that he had followed the trial, he was like a little boy feeling himself to be the victim, and denying everything: 'I didn't say that,' he'd tell the lawyers. 'I didn't do that.' 'They're bad people, making it up!'

His defence team had slowly come to wonder if his kindergarten sense of morality didn't explain his ongoing claims of innocence. Brendan had opted not to have a lawyer during his police interview, believing they were for 'bad people'. In his false Crime Stoppers report, he also referred to 'a bad man lighten fires'. Of course, good people sometimes do bad things, but to someone with lower intelligence and autism, doing a bad act makes a bad person. This concrete way of thinking is almost the opposite of defence barristers – the blurring of these concepts is the space in which they live. Ambiguity is at the heart of many cases. But Brendan, it seemed, did not want to be regarded as wicked, and therefore he would likely never change his story.

Jane Dixon, Jarrod Williams and the solicitors sat in the courtroom and waited as the jury were thanked for their service. Some of the jurors were crying, seemingly with sorrow for all involved. Then they were dismissed, and having glimpsed the complexity of other people's lives, they walked out onto the busy city street and back into their own.

Remaining in the courtroom was Kaz Sokaluk. His wife had been unable to bear these last stressful days of waiting, but he had kept travelling back and forth by train to support his son. The lawyers took in this man, already far out of his element and worn down with distress, as they all watched Brendan being shackled by security guards and escorted through the public gallery, with his lumpen, high-stepping walk, on his march to the cells.

After this, Kaz would return to Churchill and he and Lou would try to rebuild their lives. Later, the lawyers heard that one night, when it seemed things had finally become easier, the Sokaluks' house burned to the ground. Brendan's parents stood on the street watching flames from an electrical fire devour the place in which they'd raised him. It was as if some fire curse had been placed on them too.

In the courtroom, the oxygen seemed slowly to thin. The usual pleasantries the defence might have exchanged with the prosecutor were muted. A trial becomes its own little world. This one had taken priority over everything else in their lives and they'd fought and fought for Brendan, and before that worked long, hard hours in the effort to drag him through the legal process with some semblance of comprehension. Soon, exhaustion would come over all of them.

When everything was settled and they'd bidden Kaz farewell, the lawyers walked around to a grand-domed lightwell – shells and flowers in the high plasterwork, almost a kitsch vision of enlightenment. Nearby, at the foot of a Victorian spiral staircase, a wooden sign in a scrolling golden font read: NO ADMITTANCE/ PRISON STAFF ONLY.

They pressed an old-fashioned doorbell and waited to be admitted to the cells.

Up the bluestone circular steps, the architecture seemed more stifling, more draconian than ever. They'd all been in this close, grim place before, and the deja vu was perhaps part of their sudden weariness.

At the top of the stairs were the cells that seemed to have been last modified decades earlier. They consisted of a row of tiny booths with thin wooden partitions, devoid of privacy. Other barristers were there, talking to their clients through perspex barriers, and the guards, walking up and down, could hear everything.

This conversation with the freshly convicted was what some barristers considered the worst part of their job. For most people, the prospect of spending years in jail remained abstract until this point. That was a means of coping. But when they heard the guilty verdict, reality closed in. And for the defence, the reality that closed in was a feeling of responsibility, an immediate accounting of whether they'd done all that they might have, and whether the client wondered this too.

Seeing Brendan now, the usual sinking feeling was only deeper and more vertiginous, the question of what to say more difficult, since Brendan was waiting behind the scratched perspex and his facial expression and demeanour had not greatly changed. The now tear-reddened eyes and bleak looks worn by the lawyers didn't faze him, maybe because, making no eye contact, he didn't really see them. And because he didn't see them, he looked blinded and defenceless.

It was dawning on the lawyers that their client hadn't actually realised he'd been found guilty, or even that his trial was now over. The profound consequences of what had just happened in the courtroom, like the fire itself, appeared not to have registered. As Jane Dixon tried to explain the verdict to him, Brendan just went on asking when he could go home.

Later, when the verdict *had* sunk in, his mother told the lawyers that her son seemed most distressed by the prospect of missing out on years of birthday and Christmas presents. But right then, after three years of legal wrangling, and countless days in court in an atmosphere of grief and hatred, it was clear their client was as lost as ever. The legal contest had pitted the story of a fiend against that of a simpleton, but the two weren't mutually exclusive. Brendan was both things. Guileful and guileless, shrewd and naive. A man apparently capable of unleashing chaos and horror, who now, behind the perspex of the cells, looked so bewildered that when the lawyers said goodbye they felt devastated, for it seemed they were leaving behind a child.

coda

Today, in the quiet streets of Churchill, there's an abandoned brown brick house. On a hot day it looks unnaturally small, sun-blasted, like it's shrivelled in the heat. On either side, the neighbours still keep neat lawns, and hedges sheared into meticulous globes, but in the front garden of this place, behind a falling-down fence, there are only weeds. Some have seeded in the guttering and between the roof tiles. From the street, it's a few steps to the backyard, which has the brick remains of an incinerator, and a graffitied shed once used to store junk. The curtains in all the windows are drawn and signs have been pasted to the glass: PLEASE NOTE THAT ASSET CONFISCATION

OPERATIONS IS IN POSSESSION OF THIS PROPERTY PURSUANT TO THE CONFISCATION ACT 1997. Those signs are faded now. The agents of asset confiscation surely found little worth seizing. The property remains vacant.

I stand for a moment, taking it in.

It is nearly a decade after Black Saturday, and I have recently written a four-sentence letter to the house's former owner: *Dear Brendan, my name is Chloe Hooper. As you may know from talking to your lawyers, I am writing a book about the 2009 Churchill fires. If you are willing, I would like to visit you. Please let me know if I can.*

Brendan had left the Supreme Court after his guilty verdict and returned to the maximum-security prison unit with the garden program for intellectually disabled prisoners. He received a sentence of seventeen years and nine months, the judge taking into account his 'impaired mental functioning' and noting he had already served nearly three years before the trial.

Many of the victims' families were distressed at what they perceived as extreme lenience. The sentence was appealed as 'manifestly inadequate' by the Office of Public Prosecutions, but upheld. There was no dedicated program tailored to arsonists in Australian prisons, although Jane Dixon, betraying no sense of irony, revealed at the appeal hearing that her client was fulfilling a long-held ambition to undertake a horticulture course.

After five and a half years Brendan was moved to a medium-security jail with another large intellectually disabled population. This is where I sent my letter. Having spoken with the detectives and his lawyers, the latter with Brendan's and his family's permission, it seemed fair to give him the opportunity to have his say. Although 'his say', I knew, would be a fraught and partial thing.

I first visited Churchill a few days after Brendan was arrested. I don't know what I expected to see, but the Black Saturday fires had burned into my mind, and a mind that might have lit one of them seemed beyond comprehension. Blackened hills ringed the township, which, even though untouched, looked flat, luckless, in its own fever. In a crescent of near-identical houses, a group of boys rode in circles, standing wolfish on their BMX bikes, knowing something was happening, finally, in this place. They stared at the stranger who'd turned up to stare at them. At the shopping centre, too, the locals found it easy to pick the outsiders come to gawk. And there were a lot of us: the whole state was in a furore, and Churchill also heaved with emergency service workers and police.

Roads were still closed off, with barricades in every direction. I turned down one bush-lined dirt track then another. Now I didn't know where I was driving. There were no views of fire damage here. The only wrong notes were the brown paddocks, and ribbon bark peeling off the parched gums. This dryness, and the sense in the town of claustrophobia and agoraphobia com-mingling, already made the blaze seem less mysterious.

Soon, though, I realised that a station wagon was following close behind me. The driver didn't want to pass. She seemed to be a local-turned-security agent, sickened by a sightseer. She tail-gated, as if I might be planning to get out and light a copycat fire myself. After a while I did a U-turn and drove back to the city.

Now, as I stand looking at Brendan's old house, I've nearly finished writing this book, which came in fits and starts, after persuading people to speak, and learning of material that was hard to access, then too hard to deal with. I have spent years trying to understand this man and what he did, my own motivation sometimes as indecipherable as his. And, I wondered, what if, having asked the police and lawyers dozens of questions, then more questions, trying to get tiny details right, I essentially ended up with little more than a series of impressions? Would the result be ultimately a fiction?

Maybe, that morning, Brendan woke up inside this house and before long a dark idea took root. And *maybe*, by the middle of the scorching day, as he stood watching a fire truck arriving to extinguish a grassfire (the blaring siren, the flashing lights, the uniformed volunteers – a scene from the children's shows he adored, a tableau of power, adrenaline, control) the idea had grown. If he set a fire near this place where he felt inept and invisible, he could bravely fight it, or warn others they were in danger. He could punish all those bad people who thought he was an idiot *and* be their saviour . . .

And there I'd go, imagining there was a reason for an act that's senseless.

For a few weeks after sending my letter to Brendan, I checked my post office box every day in case he'd replied. Nothing. Eventually, someone who was in contact with him gave me an explanation. Brendan, who now spent his prison days making flat-pack furniture, had had trouble understanding what I'd written. He'd shown the letter to his supervisor, who told him not to write back. Even if he wished to talk, the prison would not allow it. Brendan was no longer represented by Legal Aid, but the lawyer assigned to deal with me there said she would never advise him to speak, in part due to the anxiety it could cause him. His parents apparently also wanted their son to keep his head down in jail and stay out of trouble.

I was disappointed and relieved.

If I *had* been granted official permission and ushered inside the prison to talk to Brendan, how close could any conversation have come to answering this book's central question? And even if it could be answered, would understanding why Brendan lit a fire make the next deliberate inferno any more explicable? Or preventable? I now know there isn't a standardised Arsonist. There isn't a distinct part of the brain marked by a flame. There is only the person who feels spiteful, or lonely, or anxious, or enraged, or bored, or humiliated: all the things that can set a mind – any mind – on fire.

But from another point of view, Brendan, with his morphing qualities of innocence and guile, *is* archetypal. A figure from stories long, long ago. The Greek god of fire and metalworking, Hephaestus, often represented as deformed, was given the epithets, *Kullopodíōn*, 'the halting'; *Polúmētis*, 'shrewd and crafty'; and *Amphigúeis*, 'the lame one'. Every culture has a tale about a human or an animal stealing fire, for better or for worse. In some traditions, this figure is a trickster spirit, like fire itself, a mischievous, self-absorbed, shapeshifting force both cunning and foolish, bewildered and bewildering.

In Australian Aboriginal mythology, the fire thief is often a bird. And in the Northern Territory's tropical savannahs, the 'fire hawks' – the brown falcon, and black and whistling kites – really have been seen carrying smouldering sticks in their beaks or talons to reignite and extend a fire if it's petering out. They hunt on the blaze's flanks, ready to capture small fleeing creatures. So even the birds here are fire-setters.

The point, in these mythological tales, is that fire is a magical, threatening, separate presence searching for ways to become real. The fire-lighter himself is almost irrelevant, a mere tool the element uses in what can today be understood as an ancient but intensifying pattern.

Now, year-round, at the strangest times and in the strangest places, we have a cache of tales about fire's power. As climate change extends the wildfire seasons in both hemispheres, 'mega-fires', blazes that burn more intensely and for longer, are occurring around the globe, and increasingly in places that rarely if ever usually burned, such as areas of the boreal zone. In July 2017, a fire in Greenland, suspected to be man-made, raged for weeks in a melted peat bog only sixty-five kilometres from the ice sheet. Europe had just had the worst fire season in memory, and by the year's end, hundreds of fires had also broken out across northern then southern California, even burning deep into Los Angeles.

In his essay 'The Fire Age', the ecologist and master chronicler of flame Stephen Pyne writes: 'Our pact with fire made us what we are.' It cooked our food and changed our physiology; we gathered around it for warmth and communion, we formed larger communities. We used fire to hunt, to manage the landscape, to practise agriculture. 'We hold fire as a species monopoly . . . It's our ecological signature,' claims Pyne, who continues: 'If people wanted more firepower – and it seems that most of us always do – we would have to find another source of fuel. We found it by reaching into the deep past and exhuming lithic landscapes, the fossil fallow of an industrial society.'

We started burning coal. A 'pyric transition', Pyne calls it, 'as disruptive as the coming of the aboriginal firestick . . . but it was more massive, much faster, and far more damaging':

> The new energy is rewiring the ecological circuitry of the Earth. It has scrambled ecosystems and is replacing biodiversity with a pyrodiversity – a bestiary of machines run directly or indirectly from industrial combustion. The velocity and volume of change is so great that observers have begun to speak of a new geological epoch, a successor to the Pleistocene, that they call the Anthropocene. It might equally be called the Pyrocene. The Earth is shedding its cycle of ice ages for a fire age.

Standing on Sheoke Grove, it's hot for an October day, and it's a weird heat. The sun often feels different now. It has a sharper bite. The concrete driveway here is cracked, the whole place scoured, really, by ultraviolet. I get back in my car.

At the top of the street, the tips of Hazelwood's chimney-stacks visible, I turn into Acacia Way and then Monash Way.

I pass the petrol station where Brendan Sokaluk bought a pack of cigarettes before setting the fire.

'Do you know how you light a fire and get away, you know, get right away?' he asked a neighbour in the days before his arrest.

She didn't want to know, she later told police.

He told her anyway: you put a cigarette near the fuel you want to light and escape while it's burning down to it, like a fuse.

The petrol station is shuttered now, abandoned like the brown brick house.

I drive past Eel Hole Creek, not much of a creek if it's still there at all, then I come to eucalypts in straight lines, the mono-culture of a plantation interrupted by transmission towers and powerlines, and shortly I pass a stretch of road that always brings to mind science fiction, the dystopic kind set in the concrete and rust of a post-industrial world. Behind barbed-wire fencing, various elements of power infrastructure make a tangled fortress of steel and wires, transformers and insulators, chimneys and pylons. It's a kind of grandly engineered scrapyard now. In a line are the remnants of the Morwell power station, the briquette factory, the char plant. All this industry has shut down in the past few years, one closure precipitating the next.

Further along the road is the Morwell timber mill. The gates here are also bound in chains and giant padlocks. It, too, closed a couple of months earlier. After Black Saturday, and a second bushfire five years later, the plantation ran out of wood.

I veer onto Miners Way to get on the M1 back to Melbourne, and pass various Hazelwood buildings on the outskirts of the mine.

At one stage while I was researching this book, a local woman kindly took me for a drive and we stopped on the hillside over-looking the Valley. In the midst of the rolling green pastures were the power stations with their billowing towers, emitting pollution that would be illegal in China, the United States and Europe. My guide believed that God would not have created coal if He hadn't wanted us to use it. At night, when Hazelwood was all lit up, reflected in the pondage of coolant water, it looked to her like an old paddle-steamer, oddly beautiful, even romantic.

But now this power station is closed too. More structures have been locked up, more jobs lost. This is the biggest change to hit the area during the time Brendan Sokaluk has been in jail. In early 2017, Hazelwood was unceremoniously decommissioned, regarded as an economic liability by the multinational that owned it. It's a shock to see how suddenly a building becomes a ruin. To my eye, the vast chimneystacks are reminiscent of a brutalist monument. These concrete towers have a kind of ugly grandeur and still rule the landscape as completely as they did in their heyday.

Stephen Pyne's larger point is that we have too much of the wrong kind of fire – these coal-powered flames 'stuffed' into our machines, rather than the burning regimes of Aborigines, who controlled the environment with low-intensity, mosaic blazes. This intricate knowledge that once maintained the great southern forests is now largely lost. Instead we have feral wildfires: untame-able, predatory, beastly. And we have people – sometimes raised in the social toxicity common around resource-extraction sites – who can, at a whim, bring these brutes to life.

In February 2014, five years after Black Saturday, a series of suspicious fires joined together and swept into the Hazelwood coalmine. The powerline to the water pumps soon burnt and mine workers had no firefighting capacity. The fire blazed out of control in the mine for a month, spreading poisonous smoke over the adjacent town of Morwell. Locals watched a blue haze roll down their streets and were told there were no health ramifications, even as particulate matter fluttered outside their windows. There was a spike in the area's death rate.

A government inquiry found the mine's safety standards grossly lax, leading to its current majority owner, the French energy company Engie, announcing in 2016 that they would close the power station down. The 400 million dollars required to bring it up to legal safety requirements, they claimed, made continuing to operate economically impossible. Engie estimated the cost of rehabilitating the decommissioned site at 743 million dollars: 304 million to demolish the power plant and regenerate the immediate surrounding area; 439 million to turn the mine into a massive lake, requiring more water than Sydney Harbour, with birdwatching areas, picnic facilities and walking tracks – an eerie, post-industrial version of the prehistoric swamp that started all this.

In the afternoon light, the vast inside of the mine looks golden brown. It's a pit seven and a half kilometres long, three to four kilometres wide, and roughly 150 metres deep. The sprinkler system is in operation; hundreds of arcs of water keep the coal from drying and spontaneously combusting on this hot day.

Even before the mine closed, it gave a strange sense of stasis. At least, that was the fleeting impression I would get driving by. Some button had stuck. A view of dredgers and tip trucks was on a constant loop. Nothing seemed to change, although the gouge in the ground revealed *everything* that had changed – the stratification of uncountable years. Apparently, complete fossilised tree trunks are sometimes visible within the coal. Thirty-million-year-old myrtle beech, trees from the lush Gondwanan forest predating the reign of the eucalypts. Here, the deep past feels eerily close, and if giant excavators can peel away geological time, unearthing in one movement prehistoric sediment, why can't we scratch out and reconstitute the matter of a decade? A life's most painful moments often still feel so recent you could almost reach back and undo them.

Along the road, pylons stretch out ahead, their sea of wires spreading from the Valley across the state in wavelike symmetry. On Black Saturday the majority of people died in fires caused by failures in a privatised electricity grid with inadequate safety standards and regulatory checks. Still, aging electrical infra-structure can be replaced, and monitoring these can be better regulated. In twenty or thirty years, will this even be the way electricity is delivered? These may be the feral fires we can most easily prevent.

But in places around the world with high rates of deliberately lit blazes, even if community programs are introduced to teach children fire safety and encourage locals to be vigilant, even if roads are closed and drones are used to watch for suspicious activity, even if convicted arsonists wear GPS tracking devices that bleat when they're close to national parks, fire-setting can't be stopped entirely. On hot days, police can't monitor everyone with rage inside them like a striking stone. And so, in the end, the small world of someone like Brendan is not as unconnected to our own as it might seem – and we ignore this at our peril.

If arson is an expression of a particular psychology, there will always be arsonists. And as we head towards ever warmer summers in a changing climate, there will be ever more opportunities for those drawn to lighting fires. It is increasingly fertile ground for the venting of discontent . . . Behold Brendan's inferno – his light show and proxy, all-powerful, all-destroying – converting the wounds he carried into pure pain for everyone around him. Revenge as elemental to the flames as oxygen.

I keep driving, the eucalypts on either side of the road flexing their loping limbs and biding their time.

At a small town, I make a turn. Here, there are vestiges of old Gippsland: a bakery selling neenish tarts and other bygone delicacies, a bric-a-brac and a lolly store, aging weatherboard cottages with blossom branches spilling over paling fences, a pretty church with a flagpole.

Outside the church stands a woman. She's turning a skipping rope that's tied to the pole. Lanky primary school students jump and recite the alphabet, to the *thwack* of the rope against the concrete. Round and round her strong arm rises, lifting the rope high enough to clear the children. Shirley Gibson, a petite eighty-year-old with short hair dyed a vivid burgundy, is a volunteer in an after-school program.

There are a variety of reasons why our relationship could be awkward – I'm writing, after all, about the fire that killed two of her sons – but it has not been. Shirley, an ex-English teacher now organises me into activities in the same enfolding manner she employs with the children. She encourages me to eat fruit, play board games, listen to performances of the songs she's taught on an electronic keyboard. Then, when 'her' kids have been collected, she washes up their milk-stained plastic cups and heads to her car. I follow her in mine a short distance, to a cream-brick house.

Shirley bought it from a pastor, and he must have told the townspeople her situation, because she never did and yet from the beginning the community were incredibly kind to her. A woman she'd never met, and didn't see again, turned up at the house with an enormous basket full of simple things: a hammer, pegs, even nail polish and lipstick. One of the reasons Shirley does so much volunteer work is to repay all she feels she's been given, the kindness that was so diffuse. 'Who do you go to to say thank you?' she sometimes asks.

The door is unlocked and she ushers me inside.

Her front sitting room is decorated with bouquets of flowers. They are next to photographs of David and Colin, and her late husband Bill, who died eighteen months after the fire. Bill, a boilermaker and welder who'd worked on all the Valley's power stations, had fought leukaemia for twenty-five years, but losing his sons made it harder to go on. In one frame, David is throwing his young granddaughter, who he was helping raise, into the air; in another, Colin is smiling in bemusement down the lens. Next to each picture are synthetic butterfly wings that are sent every anniversary by a support group for Black Saturday survivors.

Shirley walks through to the kitchen and checks on her pets, and of course, like the burnt seed in the sapling, or the ancient swamp in the mine, the ghost of the old house is within this one.

She can close her eyes and it's that Saturday again, the curtains drawn to keep out the heat. She's about to sit down at her sewing machine when a neighbour comes to tell them there's a fire nearby. Shirley has just unpacked a suitcase of precious things selected the week before, when the Delburn fires were lit. Now she takes only a few clothes while Bill fetches their dogs. As they are backing out of the drive, Colin arrives.

Bill asks his son to leave for his own safety; then, frightened, he becomes angry, ordering Colin to get off the property. The fire is burning less than a kilometre away. Bill is worried that David will also be coming to defend the house, and he knows Colin won't leave his brother. It isn't that they are close, they'd never really been close, but if David comes, Colin will stay. When Bill can't change his son's mind, he drives away.

Soon, David slips cold drinks to the officers manning a roadblock and is allowed onto Glendonald Road. Both brothers have been CFA volunteers at different times. They carefully prepare the house, speaking to their parents intermittently through the afternoon, discussing the location of the fire, and even laughing when their proper father tells them to get the fuck out of there. Perhaps it feels good to work together.

Colin, older by a year, is a sweet-natured man who lives on a disability pension due to a genetic condition called Klinefelter syndrome. Growing up, he was bullied mercilessly because he was clumsy, slow and extremely tall. David had been embarrassed at having an older brother whom others regarded as dumb and freakish. They were rough places, the Valley schools of the 1970s, and at that age, how could he know better? Or rather, even while knowing better, how could he resist the pull of darker instincts, the desire to fit in when fitting in feels like a matter of survival? David joined the pack and bullied his brother too.

Colin's other younger siblings worshipped their lovely, gangly-limbed brother. In the playground they fought his tormenters, and therefore they weren't close to David either. (Not that David went on to have an easy life himself. He'd raised two children on his own with what he earned driving a truck, breeding dogs, and doing security and farm work.)

How do you unmake between siblings the deep past and its complicated geology? Maybe David suffered in some way from what he'd done, and maybe Colin stayed at the house in the path of the fire that day because he wanted his brother to know he forgave him. Colin was able to grow a bigger, softer heart from what he'd had to withstand. He became a man who felt compassion for anyone who was suffering.

When Shirley and I first met she was still training herself to erase the arsonist from the fire. 'He's a *nothing*,' she told me, with bitterness that proved to be uncharacteristic. 'He doesn't exist as far as I'm concerned.' This man was her lightning strike. Her glass shard waiting in long grass.

But she immediately added that she felt sorry for Brendan Sokaluk's parents, though, because they too had lost a son. After the trial, she found herself writing a letter to Lou. In it, Shirley said that as the mother of four boys, she knew she'd had no control over any good or bad they might have done as men out in the world: she hoped this other mother did not feel responsible for her son's actions.

Shirley never received a reply. Perhaps her letter never reached the Sokaluks. Or perhaps it got lost in their grief. (Brendan's parents, it seemed, were still periodically haunted by the thought that their son was like the wrongly accused Mr Philpott, the victim of a terrible miscarriage of justice.) When, later, Shirley heard that their house had burned down, she thought of writing again, before her days filled with other things.

This evening, sitting in her chair, Shirley says that she has recently allowed herself to start thinking about Brendan. She knows he'll be out of jail before long, and to her surprise she can bear to consider his story.

She says she would like to meet him and ask why he lit the fire. The impossible question.

I remind her that he still claims to be innocent.

Shirley considers this, then says she'd prefer to think that her boys died because a man accidentally dropped a cigarette rather than deliberately set a fire.

Outside, the evening sky is lustred. Two dogs that Shirley has adopted are playing in the garden. Inside, sitting on the lid of the piano that someone has given her, are awards for community service and photos of the many children she has tutored. In a drawer are the letters of the alphabet, cut out of velvet or sandpaper, tactile aids she uses to help kids learn to read. A, B, C . . . one letter after another. Break things down to their simplest form and then go on. Try to go on. Each day. Start again and go on. For despite this new, full life, Shirley longs to go home. At night she dreams she is moving through the rooms of her old house and when she wakes it is lost once again.

afterword

The giant stone heads of Easter Island were to me, as a child in Australia, less a puzzle than a fable. This was a Brothers Grimm island of eco-cannibals – they devoured their trees, then each other – a tale almost too absurd to be a cautionary one. How could real people not notice their surrounds becoming uninhabitable to all but a thousand or so statues? The cartoon version would end with an impassive head turning and winking a great obsidian eye.

Contemporary archaeologists dispute this popular version of Easter Island's societal collapse, but these days the fable feels eerily plausible. It's happening, after all, across the Pacific on *this* island right now. In Australia's summer of fire, 18.6 million hectares of the country have already burned. That's an area two-thirds the size of England. As large as Massachusetts, Vermont and New Hampshire combined. And while it happens, our leaders – and to varying degrees the rest of us – ignore that we're filing over a cliff.

At first, our prime minister Scott Morrison reminded us there have been bushfires before. That's true. Down the decades there have been monsters – Black Thursday, Red Tuesday, Black Friday, Ash Wednesday. Some people can recite the list, but none burned on this scale or for this duration. The devastating 2009 Black Saturday fires did, however, come close to providing a vision of our warming future. On that 46°C day, hundreds of fires burned across the state of Victoria. One of them began in the Latrobe Valley, the site of the largest brown coalfields in the world. Eleven people died in this fire and a misfit named Brendan Sokaluk was soon charged with 'arson causing death'. Sokaluk had grown up locally, almost adjacent to a mine the size of a large city. In the boom years, his father worked shovelling coal at the mine's power station, Hazelwood; then, as Brendan reached adulthood and the state's power industry was privatised, a rust belt tightened around the community.

One day, driving to interview one of this fire's survivors, I found myself passing Hazelwood – the dirtiest power station in the Organisation for Economic Cooperation and Development – and beheld the behemoth afresh. It seemed a terrible irony to be writing about the desolation left by fire when a row of chimney stacks 450 feet high continued spewing out barely regulated carbon emissions. Not only did these pollutants make for a more fire-friendly atmosphere, but, as is well-documented internationally, there's a higher level of social dysfunction around resource extraction sites. In the disadvantaged Latrobe Valley, there were a disproportionate number of deliberately lit fires. The poorest communities in Australia are the ones most frequently surrounded by ash.

Researching this book, I was often struck by the way locals disregarded the billowing chimneys and coolant towers marking their horizon. To politicians living far from this blighted landscape, the Valley's various power stations were temples of prosperity and employment. (Famously, the current prime minister once flourished a lump of coal in the federal parliament, praising it as a God-given, civilising force.) But it was as if those nearby didn't actually see them anymore.

Now this seems less strange to me.

This summer I'm more aware of how we disconnect from environmental disaster.

In January, while the country burned, schoolchildren were on their extended holidays. My young sons and I visited a friend at a South Australian beach town. I over-applied sunscreen and we built sandcastles and paddled, even as we smelt the smoke from an inferno on Kangaroo Island across the water. This island, from a different paleontological storybook, was a natural Ark, preserving astonishing biodiversity. And it was being razed.

While the kids played in the dunes, I checked the news: an unspooling vision of destruction. Not just the island but the whole south-eastern seaboard of Australia was alight. Lightning and failed backburning appeared to be the main sources of ignition, but it could safely be presumed that human malefi-cence also played a role. (After Black Saturday, the Australian Institute of Criminology analysed 280,000 previous grass fires and found that 13.3% were 'maliciously' lit, and 36.6% 'suspi-ciously' lit.) By Christmas, New South Wales had been burning for four months. Fires had begun on the north coast, near Drake, an old copper mining town. The torch travelled further down the coast to the iron ore town of Port Macquarie; moving, then, in different directions, passing from forest to scrappy mining and farming places, and back again to forest, right down, 1200 miles to Victoria. People were huddled on sports fields or waiting on beaches for evacuations. And on Kangaroo Island, the army would soon start digging trenches to bury the dead wildlife, a tragic fraction of the billion Australian animals cremated in the fire. The horror was too much to take in.

But the point is, I didn't have to. The Apocalypse's frontline was elsewhere. While the flames had found city dwellers who'd dispersed on vacation, the greatest ruin was faced by Australians living in places where land is cheap: on the fringes of the bush in economically depressed, semi-rural communities. Those who can least afford to rebuild are most likely to find their houses burnt. That's the brutal sociology of fire in this country.

The arsonist may be a misfit, an outlier, even a sociopath, but he inhabits the same environment and the same community as his victims. Alienated as Brendan Sokaluk was to the point of nihilism, fire became his natural ally – the friend he never had. Together they refashioned the world as he knew it.

The area around a coalmine that has been permanently damaged environmentally, and therefore socially as well, is known as the sacrifice zone. In Australia, the perimeter of that zone keeps extending. In the last decade, it's spread to the Great Barrier Reef, which, due to unprecedented bleaching, is now two-thirds dead; to our country's main river system, in near total collapse; and to our drought-stricken food bowls. The iniquity of climate change means for now poorer Australians absorb the fall out, but the devastation is coming for us all.

In the meantime, I wonder what will be left at the end of these seemingly inextinguishable fires. It won't be stone heads – a fitting enough symbol of self-absorption – but the towers of power stations in a once-vivid land, so much of it now the colour of coal.

– Chloe Hooper, Melbourne, February 13, 2020

notes

PART I

A century ago, Henry Lawson wrote that arson expresses Lawson
wrote of fire-setting in his short story 'Crime in the Bush', *The Bulletin*,
vol. 20, 11 February 1899, reprinted in *The Macmillan Anthology of
Australian Literature*, (ed. Goodwin & Lawson), Macmillan, 1990.

Plenty of environmentalists had protested the slow death Julie Constable
gives an account of the forest's demise, in 'Reams and reams of paper:
The Strzelecki Forest campaigns' in *Earth and Industry: Stories from
Gippsland Past, Present and Future*, (ed. Eklund & Fenley), Monash
University Publishing, 2015.

This is where everything turned into milliseconds All accounts of
the fire and subsequent quoted material come directly from the
Victoria Police v. Brendan James Sokaluk brief of evidence, witness
statements from the 2009 Victorian Bushfires Royal Commission,

and Brendan Sokaluk's committal hearing and subsequent Supreme Court trial.

The Heartbreak Hills, locals once called this steep, poor country
Patrick Morgan tells this story in *The Settling of Gippsland: A Regional History*, Gippsland Municipalities Association, 1997.

Soon it would be known as Black Saturday: four hundred separate fires Statistics regarding the fires' power were provided in 2009 to the Victorian Bushfires Royal Commission by fire ecologist Dr Kevin Tolhurst.

In the mid-nineteenth century, pyromania was considered The following quote is taken from Ray, I (1838) *A Treatise on the Medical Jurisprudence of Insanity*, used by Dickins and Sugarman in their essay 'Adult Firesetters: Prevalence, Characteristics and Psychopathology', *Firesetting and Mental Health*, (ed. Dickins, Sugarman & Gannon), RCPsych, 2012.

Of vegetation fires in this country, 37 per cent The Australian Institute of Criminology analysed available data on 280 000 vegetation fires recorded by fire agencies for the 2009 Victorian Bushfires Royal Commission.

One prominent model used this equation to explain the behaviour
This model, proposed by K.R. Fineman in *American Journal of Forensic Psychology*, 13, 31–60, is quoted in 'Explanations of Firesetting: Typologies and Theories' by Theresa A. Gannon, in *The Psychology of Arson: A Practical Guide to Understanding and Managing Deliberate Firesetters*, (ed. Doley, Dickens & Gannon), Routledge, 2016.

Living back in Morwell, you were three times more likely These socio-economic and crime statistics are taken from 'Needs Assessment Snapshot: Latrobe Local Government Area, June 2016', a report by

Gippsland Primary Health Network. Morwell is included in the 2015 DOTE (Dropping of the Edge) Report as one of the country's most disadvantaged communities due to 'the degree of factors that limit people's life opportunities.' *Behind Closed Doors* by Katherine X with Sue Smethurst, Simon & Schuster, 2016, outlines a woman's long-term abuse by her father and his arrest by Morwell detectives in February 2009.

Aborigines used fire for illumination, for signalling With fire, Aborigines created the 'park-like landscape' that European sheep destroyed within a couple of generations, as Bill Gammage writes in *The Biggest Estate on Earth*, Allen & Unwin, 2011. For white settlers' experience of fire, I'm indebted to Stephen Pyne's *Burning Bush: A Fire History of Australia*, Henry Holt, 1991, and Don Watson's *The Bush*, Penguin, 2014.

PART II

McCrickard started calling the police Paul Bertoncello worked for a week asking local, interstate and international media sites to remove images of Brendan Sokaluk, with no ability to enforce his request. Victoria Police supported the subsequent suppression order.

In those days, Churchill had a population of 2500 Meredith Fletcher's *Digging People Up for Coal: A History of Yallourn*, Melbourne University Press, 1998, gives an excellent account of life in Yallourn, and the town of Churchill's beginnings.

The eight chimneystacks of Hazelwood were now *the* iconic image In November 2009 the Environmental Defender's Office announced, 'it appear[s] that total gaseous emissions from Hazelwood Power Station are not measured. Rather, emissions from one stack at a time are measured and this only occurs once every six months. Therefore, based on the annual reports submitted to the EPA which we have been provided, it appears there is limited information on the emissions of carbon monoxide, nitrogen oxides and sulfur dioxide, and on whether Hazelwood Power Station has complied with the limits on the discharge of these gases.' Many other pollutants were not measured at all. The state government's Environmental Protection Agency sold the company owning the mine, International Power, emergency approvals to pollute to protect the company from prosecution.

It wasn't surprising that many locals were distrustful of authority The Hazelwood Mine Fire Inquiry reported in 2014 on the overall health of the Latrobe Valley community, and the high rate of years lost to disease.

The commission also decided that privatised and often outdated The 2009 Victorian Bushfires Royal Commission noted that there was no way to accurately calculate the financial cost of Black Saturday, but estimated the likely sum to be 4 billion dollars.

Temple Grandin, upon seeing the joy the psychiatrist Oliver Sacks found in a sunset This episode is detailed in Sacks' *An Anthropologist on Mars*, Vintage, 1996.

PART III

When Brendan entered the dock I've drawn from Patrick Carlyon's article in the *Herald Sun* on March 20, 2012, 'Brendan Sokaluk – the boy who played with fire.'

He made it a point to never look at the accused In 2017, George Xydias wouldn't have a choice about this. An arsonist named Brendan Davies chose to defend himself in court after lighting fires in various buildings, including schools and a nightclub. During the two days he spent cross-examining Xydias, the chemist *had* to regard him. Davies, who claimed to be a 'tortured victim-creation of Australia', was autistic with a photographic memory. He had achieved notoriety after posting long YouTube videos of himself in a black balaclava, holding a cigarette lighter and ranting: 'The arsonist is using arson to strike and hit his society back as a form of justice, of vengeance ... arsonists have every right to attack society because society has done everything wrong to them.'

CODA

In the midst of the rolling green pastures were the power stations
In the article 'Coal-fired power stations caused surge in airborne mercury pollution, study finds', *The Guardian*, 4 April 2018, a report from Environmental Justice Australia is cited that claims 'Airborne mercury pollution from coal-fired power stations in Victoria's Latrobe Valley increased 37% in just 12 months.'

acknowledgements

This book would not have been possible without the support and patience of Paul Bertoncello, and his colleagues at Victoria Police. I'd also particularly like to thank Selena McCrickard for her trust, and her colleagues at Victoria Legal Aid, including Dierdre McCann, for their time and candour. I'm truly grateful to Ray Elston, and, of course, to Shirley Gibson. Thank you to everyone else who generously spoke to me about Black Saturday on and off the record.

For their help and advice, my thanks to Jaye Kranz, Tracy Bohan, Andrew Wylie, Paul Read, Janet Stanley, Amit Lotan, Patrick Kennedy, Rachel Nolan, Simon Gatt, Paul Reynolds, Brett Kahan, Patrick and Anne Morgan, David Sexton, Julian McMahon, Tom Gyorffy, Louisa Maggio, Jovelyn Barrion, Lucy Kostos, and *especially* the T and J Hooper Child Diversion Services. I also gratefully acknowledge the very generous support of a Sidney Myer Creative Fellowship.

Working with Ben Ball and Meredith Rose has been a gift of immeasurable value, and their talent, dedication and judgement truly inspiring.

Thank you to Rachel Scully for graciously stepping in and giving support in this book's last stages.

Finally, my love and thanks to Don Watson.